Mathematical Methods for Cryocoolers

Online at: https://doi.org/10.1088/978-0-7503-4826-3

Mathematical Methods for Cryocoolers

Hannah Rana
Center for Astrophysics | Harvard & Smithsonian, Cambridge, MA, USA

IOP Publishing, Bristol, UK

ISBN 978-0-7503-4826-3 (ebook)
ISBN 978-0-7503-4824-9 (print)
ISBN 978-0-7503-4827-0 (myPrint)
ISBN 978-0-7503-4825-6 (mobi)

DOI 10.1088/978-0-7503-4826-3

Version: 20251201

IOP ebooks

British Library Cataloguing-in-Publication Data: A catalogue record for this book is available from the British Library.

Published by IOP Publishing, wholly owned by The Institute of Physics, London

IOP Publishing, No.2 The Distillery, Glassfields, Avon Street, Bristol, BS2 0GR, UK

US Office: IOP Publishing, Inc., 190 North Independence Mall West, Suite 601, Philadelphia, PA 19106, USA

In loving memory of my father—your love my constant term;
your encouragement my guiding vector.

Contents

Author biography

Hannah Rana

Hannah Rana is a Clay Fellow at the Center for Astrophysics | Harvard & Smithsonian and the Black Hole Initiative at Harvard University. She works on the Black Hole Explorer (BHEX) space mission concept and is the Cryogenics Co-Lead for the BHEX instrument. Prior to this, Hannah was a Schmidt Science Fellow, also at Harvard, and has held previous appointments at the NASA Jet Propulsion Laboratory, the California Institute of Technology, the European Space Agency, and CERN. She completed her DPhil at the University of Oxford. Over the years, her research has focused on astrophysics instrumentation, black hole science, thermodynamics, and mathematical modelling.

Symbols

A	Cross-sectional area (m^2)
a_n	Coefficients in general expansion series
A_{osc}	Amplitude of oscillation (m)
α	Thermal diffusivity ($m^2\,s^{-1}$); also learning rate in regression
b_n	Coefficients in Fourier series
c	Specific heat capacity ($J\,kg^{-1}\,K^{-1}$)
c_d	Damping coefficient ($kg\,s^{-1}$)
c_n	Coefficient in complex exponential expansion
c_p	Specific heat at constant pressure ($J\,kg^{-1}\,K^{-1}$)
c_v	Specific heat at constant volume ($J\,kg^{-1}\,K^{-1}$)
C	Heat capacity ($J\,K^{-1}$)
CFL	Courant–Friedrichs–Lewy number (–)
δ	Small perturbation
δ_b	Boundary layer thickness
δ_ν	Viscous penetration depth (m)
δ_κ	Thermal penetration depth (m)
ΔT	Temperature difference (K)
Δt	Time step (s)
Δx	Spatial step size (m)
D	Hydraulic diameter (m)
E	Energy (J)
ϵ	Numerical error or residual
f	Frequency (Hz)
$f(x)$	Generic function in expansion or analysis
F	Force (N)
Fo	Fourier number, $\alpha\Delta t/\Delta x^2$
γ	Ratio of specific heats c_p/c_v
ϕ	Phase angle
ϕ_f	Scalar field variable
H	Enthalpy (J)
i	Imaginary unit, $\sqrt{-1}$
k	Thermal conductivity ($W\,m^{-1}\,K^{-1}$)
k_s	Spring stiffness ($N\,m^{-1}$)
k_w	Wavenumber ($rad\,m^{-1}$)
L	Characteristic or physical length (m)
m	Mass (kg)
$\dot{m}$	Mass flow rate ($kg\,s^{-1}$)
n	Summation index or polytropic exponent
N	Number of spatial grid points
p	Pressure (Pa)
p'	Pressure fluctuation (Pa)
P	Power (W)
Pr	Prandtl number (–)
Q	Heat or energy transfer (J)
$\dot{Q}$	Heat transfer rate (W)
Q_{ac}	Acoustic power (W)
R	Ideal gas constant ($8.314\,J\,mol^{-1}\,K^{-1}$)

R^2	Coefficient of determination in regression
Re	Reynolds number (–)
ρ	Density ($\mathrm{kg\,m^{-3}}$)
s	Entropy per unit mass ($\mathrm{J\,kg^{-1}\,K^{-1}}$)
S	Entropy ($\mathrm{J\,K^{-1}}$)
σ	Standard deviation
σ_{p}	Porosity
t	Time (s)
T	Temperature (K)
T_{H}	Hot reservoir temperature (K)
T_{L}	Cold reservoir temperature (K)
τ	Tortuosity factor
u	Particle or fluid velocity ($\mathrm{m\,s^{-1}}$)
U	Internal energy (J)
V	Volume ($\mathrm{m^3}$)
W	Work (J)
$\dot{W}$	Power (W)
x	Spatial coordinate (m)
$x(t)$	Time-domain signal
X	Complex phasor amplitude
$X(f)$	Frequency-domain representation of signal
y	Displacement or dependent variable
ω	Angular frequency ($\mathrm{rad\,s^{-1}}$)
ω_0	Natural angular frequency ($\mathrm{rad\,s^{-1}}$)
ω_{d}	Damped angular frequency ($\mathrm{rad\,s^{-1}}$)

Chapter 1

Cryocooler fundamentals

1.1 Background

1.1.1 Historical origins of cryogenics

Cryogenics, the science of achieving and maintaining temperatures below approximately 120 K, has evolved over the past 150 years into a cornerstone of modern physics and engineering. Its progression from the liquefaction of gases to the highly engineered cryocoolers aboard space observatories reflects humanity's relentless pursuit of lower temperatures, deeper understanding of matter, and technological control over the extreme.

The birth of cryogenics can be traced to the late nineteenth century, during a period of intense experimentation in thermodynamics and gas behaviour. The fundamental One key mechanism in many liquefaction systems is the Joule–Thomson effect, whereby a gas cools upon expansion at constant enthalpy if it is below its inversion temperature. Other schemes also use isentropic turbine expansion (Claude/Brayton cycles).

In 1877, two European scientists, Louis Paul Cailletet in France and Raoul Pictet in Switzerland, independently succeeded in liquefying oxygen. Cailletet used a rapid expansion technique involving compressed oxygen gas in a sealed tube, which cooled upon expansion due to the Joule–Thomson effect. Pictet, meanwhile, developed a cascade process involving sulphur dioxide and carbon dioxide to pre-cool oxygen, allowing it to condense at high pressure (Barron 1985). These breakthroughs disproved the long-standing belief that some gases could not be liquefied, ushering in an era of rapid experimentation.

The race to liquefy lighter gases continued. In 1898, James Dewar, already known for his invention of the vacuum-insulated Dewar flask, achieved the liquefaction of hydrogen at 20.3 K using a cascade process and regenerative cooling. His work also demonstrated the limitations of traditional materials at cryogenic temperatures, as many seals, insulators, and containers became brittle or ineffective.

doi:10.1088/978-0-7503-4826-3ch1

The final major challenge was helium, with a boiling point of 4.2 K. Heike Kamerlingh Onnes at the University of Leiden, after years of systematic investigation, successfully liquefied helium in 1908 using pre-cooled helium gas and a Joule–Thomson expansion circuit (Onnes 1908). His laboratory became a global hub for low temperature research and would eventually lead to the discovery of a range of novel physical phenomena.

1.1.2 The scientific foundations of cryogenics

The liquefaction of helium opened the door to entirely new physical regimes. Onnes' observation in 1911 that mercury lost all electrical resistance at 4.2 K marked the discovery of superconductivity (Onnes 1911). This unexpected behaviour revealed that entirely new phases of matter emerge at cryogenic temperatures.

In the following decades, researchers uncovered superfluidity in liquid helium-4 in 1937 and helium-3 in 1972, quantum hydrodynamics, and Bose–Einstein condensation (BEC). Each of these discoveries was made possible by increasingly sophisticated cryogenic platforms and thermometry. Cryogenic systems enabled the construction of dilution refrigerators and helium-3/helium-4 mixing circuits that achieved millikelvin and eventually microkelvin regimes.

The field also found industrial applications. Liquid oxygen became essential in metallurgy and chemical synthesis. Liquid hydrogen powered the upper stages of space rockets. Vacuum-insulated transfer lines and cryostats emerged to safely store and transport these fluids.

1.1.3 The rise of mechanical cryocoolers

Until the mid-twentieth century, low temperatures were achieved primarily using stored cryogens. However, the advent of spaceflight, battlefield electronics, and long duration physics experiments demanded autonomous, closed cycle systems that could cool reliably without the need for frequent refilling.

The Gifford–McMahon (GM) cryocooler, invented in the late 1950s and refined in the early 1960s, introduced a new approach using helium gas as a working fluid. The cycle involved alternating compression and expansion of helium gas through a regenerator packed with mesh or rare-earth materials to facilitate heat exchange (Gifford and McMahon 1960). The system included a rotary valve to control flow phases and could reach temperatures below 10 K, suitable for superconducting magnets and infrared sensors.

In parallel, Stirling cryocoolers were developed. Based on Robert Stirling's 1816 heat engine cycle, these coolers operated with isothermal compression and expansion, and included a regenerator that greatly improved efficiency. Stirling systems became favoured in defence and aerospace because of their compactness and ability to be miniaturised.

1.1.4 The pulse tube revolution

A major innovation came in 1963 with the development of the pulse tube cryocooler (PTC) and was first published in 1964 by Gifford and Longsworth Gifford and

Longsworth (1964). This system replaced moving pistons at the cold end with oscillating gas pressure waves. A closed tube with one end at room temperature and the other attached to a regenerator allows acoustic waves to transfer heat across the system.

Early pulse tubes suffered from poor phase relationships between pressure and mass flow, which limited their efficiency. However, by the 1980s, several innovations emerged that dramatically improved performance. Inertance tubes were introduced as thin, long tubes added to the warm end of the pulse tube to introduce a phase lag. Double-inlet designs enhanced phase control by introducing flow resistance. Active phase shifters, using electronically controlled components, enabled fine tuned resonance.

These improvements led to pulse tube coolers that could rival Stirling coolers in efficiency, and surpass them in reliability. The lack of moving parts at the cold tip drastically reduced vibration and increased lifespan, making them especially well suited for spaceborne and quantum systems.

1.1.5 Cryocoolers in space

Cryocoolers have become indispensable in space science. Missions such as the James Webb Space Telescope (JWST), Planck, and JAXA's JEM/SMILES pioneered advances in the use of low temperature mechanical cooling for space science purposes. The Planck spacecraft used passive radiators, a $\sim$20 K H_2 sorption cooler, a 4 K He-4 Joulse-Thomson cooler, and a 0.1 K He-3/He-4 dilution refrigerator (Triqueneaux *et al* 2006). The JWST employs a hybrid system using a pulse tube pre-cooler and a helium JT loop to cool its MIRI instrument to 6 K (Penanen *et al* 2022), with a design life of $\sim$10 years.

Cryocoolers for space missions must meet strict constraints. These include total mass budgets often <30 kg and mean time to failure (MTBF) on the order of 50,000–100,000 h, mission-dependent; extremely low vibration to avoid disturbing sensitive optics, and autonomous control without human intervention. The development of such coolers represents the culmination of decades of innovation across thermodynamics, materials science, and systems engineering.

1.1.6 The future of cryocoolers

Modern advancements target miniaturisation, efficiency, and intelligent control. Additive manufacturing is enabling more compact and intricate regenerators and heat exchangers. Magnetically coupled displacers are being developed to eliminate mechanical seals and reduce leakage. Machine learning techniques are being integrated into fault prediction systems to extend operational life, particularly for long duration missions. Hybrid systems that integrate Stirling, pulse tube, and JT components in a single architecture are also being pursued.

Ongoing research aims to push the boundaries of low vibration sub-kelvin cooling for quantum computing, gravitational wave detectors, and other frontier technologies.

1.1.7 Scientific and engineering applications

Cryocoolers have found indispensable utility across a spectrum of scientific and industrial domains. In space-based astronomy, as discussed in section 1.1.6, they are critical for the cooling of infrared and submillimeter detectors to suppress thermal background noise.

In quantum computing and condensed matter physics, cryocoolers serve as pre-cooling stages for dilution refrigerators that operate at millikelvin temperatures. Quantum bit coherence times, superconducting transition fidelity, and Josephson junction stability all benefit from reduced phonon interactions at cryogenic temperatures (Devoret and Schoelkopf 2013, Krinner *et al* 2019). ADRs are especially useful for space observatories requiring sub-kelvin stability over long durations.

In particle physics, cryocoolers are used to operate calorimeters, silicon drift detectors, and superconducting magnets. The ATLAS and CMS detectors at the Large Hadron Collider rely heavily on cryogenic systems to maintain superconductivity in their magnetic components (CERN 2008).

In medical imaging, GM cryocoolers have increasingly replaced traditional liquid helium systems in magnetic resonance imaging (MRI) scanners, reducing operating costs and environmental hazards associated with helium boil-off (Barron 1985). Emerging fields such as quantum communication, low temperature metrology, and cryogenic power electronics continue to expand the role of cryocoolers. In aerospace, cryogenic propulsion and fuel storage systems for lunar and Mars missions rely on cryocoolers to maintain liquid hydrogen and methane at cryogenic temperatures over long durations (Nugent *et al* 2024).

In industry, cryocoolers support gas liquefaction, cryopumping in vacuum systems, and superconducting magnetic energy storage. The diversity of their operational contexts demands a correspondingly diverse mathematical modelling toolkit, tailored to the specific thermal, mechanical, and fluid dynamic conditions of each system.

1.2 Types of cryocoolers

Cryocoolers can be classified based on their thermodynamic cycles, working principles, and mechanical architectures. This section presents a detailed overview of the major types of cryocoolers, highlighting their physical operation, research frontiers, and application domains.

1.2.1 Stirling cryocoolers

Stirling cryocoolers operate based on the Stirling thermodynamic cycle, which consists of two isothermal and two isochoric processes. This cycle is realised using a closed regenerative system in which the working fluid, commonly helium, is compressed and expanded cyclically between the hot and cold ends. During the isothermal compression phase, heat is rejected to the surroundings, whereas in the isothermal expansion phase, heat is absorbed from the region to be cooled.

A displacer piston facilitates the transfer of gas between the two ends of the system. The gas passes through a regenerator, typically composed of fine wire mesh, metal spheres, or rare-earth materials. This regenerator plays a pivotal role in storing and releasing thermal energy, thereby enhancing the overall cycle efficiency by reducing entropy generation between strokes (Walker 1983). High performance Stirling systems often incorporate flexure-bearing, spring-suspended pistons, which reduce friction, suppress wear, and ensure hermetic operation. These design elements contribute to the long term reliability and low maintenance profile of sealed Stirling units.

In terms of ongoing research, several avenues are under active development. These include optimisation of regenerator packing materials to improve heat exchange while minimising pressure drop, as well as vibration cancellation strategies to reduce exported mechanical disturbance. Another area of advancement involves reducing parasitic heat loads and thermal losses by improving insulation and compact thermal architecture. Active control of displacer motion using linear motors and feedback control loops is also gaining attention to fine tune performance under varying thermal loads.

Stirling cryocoolers have been deployed in a wide range of applications. They have seen use in Earth-observing satellites such as the RHESSI mission, in which they cooled germanium detectors. They are also commonly used in military thermal imaging systems and portable cryogenic cameras. In addition, Stirling coolers are often employed as pre-cooling stages in hybrid systems that combine different thermodynamic cycles for enhanced temperature reach or cooling capacity (Radebaugh 2009).

1.2.2 Pulse tube cryocoolers

Pulse tube cryocoolers (PTCs) are regenerative systems that utilise oscillatory pressure waves rather than mechanical pistons at the cold end. The absence of moving parts in the low temperature region is a major advantage, as it significantly reduces mechanical wear and eliminates vibrational noise, making PTCs especially suitable for sensitive instrumentation.

A typical PTC operates by generating sinusoidal pressure oscillations using a compressor, which drives helium gas through a regenerator into a pulse tube. The regenerator preconditions the gas thermally, allowing the gas to arrive at the pulse tube with controlled entropy content. A phase shifter, such as an inertance tube connected to a reservoir, introduces a time delay between pressure and flow, establishing the conditions required for net enthalpy transport. As a result, heat is absorbed at the cold end and rejected at the warm end, enabling refrigeration (Radebaugh 2000).

Advancements in pulse tube technology over the last three decades have focused on optimising phase shifter geometry, improving regenerator packing, and adopting active control electronics for adaptive resonance tracking. Research has also explored multi-stage PTCs that extend cooling capabilities to below 4 K. These multi-stage configurations rely on cascading pulse tubes with individual regenerators and phase shifting elements tuned for each stage.

Pulse tube cryocoolers are widely used in space-based detectors. For instance, they are utilised in transition-edge...... utilised in transition-edge sensor (TES) based x-ray instruments, superconducting quantum interference devices (SQUIDs), and superconducting optical resonators (Radebaugh 2009). Their minimal vibration footprint makes them highly favourable for interferometry and precision timing applications.

1.2.3 Gifford–McMahon cryocoolers

The GM cryocooler is another type of regenerative cryocooler, originally developed in the late 1950s. It resembles the Stirling cycle in basic structure but implements discrete pressure modulation using a rotary valve rather than a continuous sinusoidal input. This modulation allows helium gas to undergo compression and expansion phases as it flows through a regenerator and into a cold end expansion volume (Gifford and McMahon 1960).

The displacer in a GM cooler is mechanically driven and used to transfer gas back and forth, enhancing thermal contact with the regenerator. Although the mechanical complexity is greater than in a PTC, the GM cooler offers robustness and simplicity in control. GM systems generally exhibit higher levels of mechanical vibration and are more massive than their Stirling and pulse tube counterparts. However, they compensate with high reliability, high cooling power at modest cost, and operational stability.

Research in GM cryocoolers is currently focused on improving efficiency, reducing valve leakage, and lowering vibration. Efforts are also directed toward miniaturising the systems for compact superconducting magnet cooling and integrating multi-stage GM heads to reach temperatures of 2–3 K.

GM cryocoolers are widely used in commercial and scientific applications. These include MRI scanners, cryopumps used in semiconductor fabrication, superconducting magnets, and laboratory cryogenics for material characterisation and low temperature physics (Barron 1985).

1.2.4 Joule–Thomson cryocoolers

JT cryocoolers operate on the principle of gas expansion through a throttling valve or orifice. The working gas, typically helium or nitrogen, is pre-cooled before it expands through the restriction, leading to a temperature drop due to the Joule–Thomson effect. Since helium has a relatively low inversion temperature of approximately 40 K, JT systems require an upstream pre-cooling stage, such as a Stirling or pulse tube cooler, to bring the gas below this threshold before effective cooling can occur (Radebaugh 2009).

The heat exchange between the incoming high pressure gas and the outgoing cold gas is facilitated using compact counterflow heat exchangers. These components recover enthalpy and improve system efficiency. Cooling capacity and performance are governed by several factors, including inlet pressure, gas flow rate, valve geometry, and the thermal efficiency of the pre-cooling stage.

Research efforts are concentrated on optimising the miniaturisation and integration of JT systems, particularly for embedded and aerospace applications. This includes the development of microchannel heat exchangers and oil-free compressors for improved long term reliability. Another focus is the use of alternative gas mixtures to tailor cooling profiles for specific detector types or payload thermal requirements.

JT cryocoolers have seen extensive use in aerospace, particularly for cooling detectors on instruments such as the Mid-Infrared Instrument (MIRI) aboard the JWST. They are also employed in cryogenic storage tanks, low noise telecommunications lasers, and other systems where passive operation, compactness, and fault tolerance are paramount (Ross 2022).

1.2.5 Turbo–Brayton cryocoolers

Turbo–Brayton cryocoolers implement a reverse Brayton cycle using turbomachinery to compress and expand the working gas in a continuous flow system. The gas, typically helium or nitrogen, is compressed (approximately) isentropically and passed through a recuperative heat exchanger before being expanded in the ideal reverse-Brayton cycle in a turbine, resulting in cooling.

Because the expansion occurs through a high speed turbine rather than a throttle valve, the process is nearly isentropic and therefore more efficient than the Joule–Thomson process. The recuperative heat exchanger plays a crucial role in exchanging energy between the warm return stream and the incoming high pressure gas, thus enhancing overall thermodynamic performance.

Modern Turbo–Brayton systems operate at tens of thousands of revolutions per minute and require sophisticated electronic controllers for startup, balancing, and thermal regulation. They tend to be bulkier than regenerative systems but offer the advantages of extremely low vibration, continuous flow, and scalability to high cooling loads.

Active research areas include the development of oil-free bearing systems, high efficiency micro-turbines, and integration with hydrogen or methane storage tanks for space propulsion systems. In addition, hybrid Brayton–JT or Brayton–Stirling systems are being explored for applications requiring both deep cooling and high thermal lift.

These systems are particularly attractive for long duration missions that require continuous-duty refrigeration, such as cryogenic depots in orbit, spacecraft heat rejection platforms, and large telescope thermal shielding (Nugent *et al* 2024).

1.2.6 Adiabatic demagnetisation refrigerators

Adiabatic demagnetisation refrigerators (ADRs) rely on the magnetocaloric effect to achieve sub-kelvin temperatures. The working medium is a paramagnetic salt, typically ferric ammonium alum (FAA) or cerium magnesium nitrate (CMN), which exhibits entropy reduction upon magnetisation. When exposed to a magnetic field under isothermal conditions, the magnetic moments align and reduce entropy. Subsequently, during adiabatic demagnetisation, the process is (approximately) isentropic; the temperature drops as the field is reduced.

A heat switch is employed to thermally decouple the ADR stage during the demagnetisation phase, thereby preserving the low temperature. These systems can be operated cyclically or in quasi-continuous mode using multiple staggered ADR stages. Contemporary ADRs incorporate superconducting solenoids, gas-gap heat switches, and optimised salt compositions for extended hold times and rapid cycling.

Research into ADRs centres on improving cycle efficiency, reducing parasitic heat loads, and developing advanced materials with high magnetocaloric coefficients. ADRs are also being combined with pulse tube and JT coolers to create composite sub-kelvin systems for highly sensitive instruments.

These devices have been used in several astrophysical missions, including NASA's Astro-E2 and Hitomi, to cool transition-edge sensors and x-ray calorimeters. They are also employed in dark matter detectors, quantum computing testbeds, and ultra-low-noise photon detection systems (Shirron *et al* 2004).

1.3 Book layout

This book is structured to provide a rigorous yet accessible treatment of the mathematical methods underpinning cryocooler design, analysis, and optimisation. Each chapter builds upon the preceding content, guiding the reader from foundational principles through to advanced modelling techniques. The overall progression reflects both the physical architecture of cryogenic systems and the computational strategies required to characterise their performance.

Chapter 2 introduces the core mathematical tools that are frequently employed in cryocooler analysis. It covers topics such as complex numbers, expansion series, and differential equations, laying the groundwork for subsequent modelling frameworks.

Chapter 3 presents a detailed treatment of thermodynamic cycles relevant to cryogenic refrigeration, including the Stirling, Gifford–McMahon, Joule–Thomson, and pulse tube cycles. The focus is on pressure–volume and temperature–entropy representations, exergy analysis, and cycle optimisation strategies.

Chapter 4 explores harmonic approximations, introducing the concept of phasor analysis for oscillatory cryocooler components. This provides a simplified yet powerful framework for modelling steady-periodic flow and heat transfer behaviour in systems driven by sinusoidal inputs.

Chapter 5 addresses numerical methods for cryocooler modelling. It introduces one-dimensional and multidimensional finite difference and finite volume approaches, including time-stepping strategies, stability analysis, and benchmarking. Comparisons are made between analytical, phasor-based, and time-domain numerical results.

Chapter 6 develops the theory of thermoacoustics, with a focus on Rott's equations and boundary layer analysis. This chapter forms the link between frequency-domain representations and distributed field models of acoustic power transport and dissipation.

Chapter 7 delves into computational fluid dynamics (CFD) and hydrodynamic modelling for cryocoolers. It discusses flow separation, porous media effects, low temperature non-equilibrium (LTNE) models, and the role of turbulence in regenerator and pulse tube behaviour.

Chapter 8 is dedicated to signal analysis and optimisation methods. It introduces Fourier transforms, spectral analysis, and machine learning-based approaches to performance prediction, fault detection, and design optimisation in cryogenic systems.

Throughout the book, emphasis is placed on physical insight, mathematical clarity, and practical application. Worked examples and case studies are provided where possible to contextualise the techniques. The aim is to equip engineers, physicists, and applied mathematicians with the tools needed to model, interpret, and innovate within the field of cryogenic engineering.

References

Barron R F 1985 *Cryogenic Systems* (Oxford: Oxford University Press)

CERN 2008 *Cryogenic heat load and refrigeration capacity management at the Large Hadron Collider (LHC)* LHC Project Report 1171 CERN. https://cds.cern.ch/record/1173062/files/LHC-PROJECT-REPORT-1171.pdf

Devoret M and Schoelkopf R 2013 Superconducting circuits for quantum information: an outlook *Science* **339** 1169–74

Gifford W E and Longsworth R C 1964 Pulse tube refrigeration *J. Eng. Ind.* **86** 264–8

Gifford W E and McMahon H D 1960 Regenerative cryogenic cycle *US Patent* USA 3,218,815

Kamerlingh Onnes H 1908 Further experiments with liquid helium *Leiden Commun.* **124c** 818–21

Kamerlingh Onnes H 1911 The superconductivity of mercury *Comm. Phys. Lab. Univ., Leiden* **120b** 122–4

Krinner S *et al* 2019 Engineering cryogenic setups for 100-qubit scale superconducting circuits *EPJ Quantum Technol.* **6** 2

Triqueneaux S *et al* 2006 Design and performance of the dilution cooler system for the Planck mission *Cryogenics* **46** 288–97

Nugent B T *et al* 2024 Cryocooler technology opportunities within space exploration *Cryocoolers 23, International Cryocooler Conference (Madison, WI)*

Penanen K *et al* 2022 Mid-infrared instrument cryocooler on James Webb Space Telescope: Cooldown, commissioning, and initial performance *Cryocoolers 22, International Cryocoolers Conference (Boulder, CO)*

Radebaugh R 2000 Pulse tube cryocoolers for cooling infrared sensors *Proc. SPIE* **4130** 363–79

Radebaugh R 2009 Cryocoolers: the state of the art and recent developments *J. Phys.: Condens. Matter* **21** 164219

Rana H 2021 Stirling pulse tube cryocoolers for space science missions *DPhil Thesis* University of Oxford

Rana H 2023 *Advancing cryogenic systems for the next generation of astrophysics discoveries, IAF Technical Papers* No. 74

Ross R G 2022 Conceptual design and development history of the MIRI cryocooler system on JWST *Cryocoolers 22, International Cryocoolers Conference (Boulder, CO)*

Shirron P *et al* 2004 Development of a cryogen-free continuous ADR for the constellation-X mission *Cryogenics* **44** 581–8

Walker G 1983 *Cryocoolers, Part 1 and 2* (Berlin: Springer)

Chapter 2

Mathematical fundamentals

2.1 Complex numbers

A complex number is a number that comprises two components: a real part and an imaginary part. These numbers extend the idea of the one-dimensional number line to the two-dimensional complex plane by introducing a new dimension perpendicular to the real axis. The key to this extension is the definition of the imaginary unit, denoted by i, which is defined as the square root of -1:

$$\mathrm{i} = \sqrt{-1}. \tag{2.1}$$

This definition does not correspond to any real number, since no real number squared yields a negative result. Nevertheless, by postulating the existence of i, we open the door to a richer number system known as the complex numbers.

Using the imaginary unit, we define a complex number z as an ordered pair of real numbers (a, b), which is conventionally written in the algebraic form

$$z = a + b\mathrm{i}, \tag{2.2}$$

where a and b are real numbers. Here, a is the *real part* of z, denoted $\Re(z)$, and b is the *imaginary part* of z, denoted $\Im(z)$.

The term 'imaginary' can be misleading—although historically controversial, imaginary numbers have very concrete interpretations and essential applications in mathematics, physics, and engineering. The combination of a real part and an imaginary part allows complex numbers to represent a much wider variety of phenomena than real numbers alone. For example, they are essential in solving polynomial equations that have no real solutions, such as $x^2 + 1 = 0$, whose solutions are $x = \pm\mathrm{i}$. Complex numbers also provide a natural language for describing oscillatory behaviour, wave propagation, and transformations in two-dimensional space.

Graphically, complex numbers can be visualised on the complex plane, with the horizontal axis representing the real part and the vertical axis representing the

doi:10.1088/978-0-7503-4826-3ch2

imaginary part. In this interpretation, the number $z = a + bi$ corresponds to the point (a, b) in the plane, allowing us to understand complex arithmetic in geometric terms as well as algebraic.

2.1.1 Basic operations

Understanding how to manipulate complex numbers is fundamental to mastering their application in mathematics and engineering. Each of the basic arithmetic operations—addition, subtraction, multiplication, and division—can be performed using algebraic rules analogous to those for real numbers, but with special attention to the imaginary unit i.

Given two complex numbers $z_1 = a + bi$ and $z_2 = c + di$ we have the following operations.

2.1.1.1 Addition and subtraction

Complex addition and subtraction are performed component-wise, just as with vectors. That is, the real parts are added or subtracted separately from the imaginary parts:

$$\begin{aligned} z_1 + z_2 &= (a + c) + (b + d)\mathrm{i}, \\ z_1 - z_2 &= (a - c) + (b - d)\mathrm{i}. \end{aligned}$$

These operations can be visualised geometrically in the complex plane. When adding two complex numbers, we treat them as vectors and perform vector addition. The result corresponds to the diagonal of the parallelogram formed by placing the vectors tip-to-tail. Subtraction can be interpreted as the vector from the tip of z_2 to the tip of z_1.

2.1.1.2 Multiplication

Multiplying complex numbers is slightly more involved but follows distributive multiplication:

$$\begin{aligned} z_1 z_2 &= (a + b\mathrm{i})(c + d\mathrm{i}) \\ &= ac + ad\mathrm{i} + bc\mathrm{i} + bd\mathrm{i}^2 \\ &= (ac - bd) + (ad + bc)\mathrm{i}. \end{aligned}$$

Here, we use the identity $\mathrm{i}^2 = -1$ to simplify the expression. This operation combines both a scaling (by the moduli) and a rotation (by the arguments) of complex numbers in the complex plane. Multiplication rotates one number by the angle of the other and scales its length accordingly.

2.1.1.3 Conjugate and modulus

The *complex conjugate* of a complex number $z = a + bi$ is the number $\bar{z} = a - bi$, which has the same real part but an imaginary part of opposite sign. Graphically, taking the conjugate corresponds to reflecting the complex number across the real axis in the complex plane.

The *modulus* of a complex number, denoted $|z|$, is the Euclidean distance of the point z from the origin in the complex plane:

$$|z| = \sqrt{a^2 + b^2}. \tag{2.3}$$

This is analogous to the magnitude of a vector. The modulus plays an essential role in the polar and exponential representations of complex numbers and in assessing their size or strength in applications.

2.1.1.4 Division

Division of complex numbers is performed by multiplying the numerator and the denominator by the complex conjugate of the denominator. This eliminates the imaginary part in the denominator:

$$\frac{z_1}{z_2} = \frac{a + bi}{c + di} = \frac{(a + bi)(c - di)}{(c + di)(c - di)} = \frac{(a + bi)(c - di)}{c^2 + d^2}. \tag{2.4}$$

This technique is known as 'rationalising the denominator'. The denominator becomes a real number $c^2 + d^2$, which makes the result easier to interpret and calculate. The numerator can be expanded using the multiplication rule, and the resulting expression is separated into real and imaginary parts. This approach ensures the result remains in the standard form $x + yi$.

2.1.2 Polar form of complex numbers

Although the rectangular (or Cartesian) form $z = a + bi$ is algebraically convenient, many applications, particularly those involving multiplication, division, or exponentiation, are more naturally handled using the polar form of complex numbers.

The polar form expresses a complex number in terms of its distance from the origin (modulus) and the angle it makes with the positive real axis (argument). For a complex number $z = a + bi$, we define:

- $r = |z| = \sqrt{a^2 + b^2}$: the **modulus** or magnitude of the complex number,
- $\theta = \arg(z)$: the **argument** or angle (in radians) the complex number makes with the real axis.

Using trigonometric relationships in the complex plane, we can write the complex number as

$$z = r(\cos\theta + \mathrm{i}\sin\theta). \tag{2.5}$$

This is known as the **polar form** of a complex number.

The angle θ can be determined using the inverse tangent function:

$$\theta = \tan^{-1}\left(\frac{b}{a}\right). \tag{2.6}$$

However, care must be taken with the quadrant in which the complex number lies. Many computing environments provide a function such as `atan2`(b, a) to correctly compute the angle θ over the full $[-\pi, \pi]$ or $[0, 2\pi]$ range.

2.1.2.1 Geometric interpretation

One of the most powerful ways to understand and visualise complex numbers is through their representation on the complex plane, also known as the Argand diagram. In this two-dimensional coordinate system, the horizontal axis represents the real part of a complex number, and the vertical axis represents the imaginary part. A complex number $z = a + b\mathrm{i}$ is thus plotted as the point (a, b), where $a = \Re(z)$ and $b = \Im(z)$.

This geometric representation allows us to treat complex numbers as vectors in the plane, providing a visual and intuitive understanding of complex arithmetic. Operations such as addition, multiplication, and conjugation take on elegant geometric meanings when interpreted in this context.

The geometric interpretation of a complex number in polar form is elegantly illustrated in the Argand diagram shown in figure 2.1. In this diagram, the complex number $z = a + b\mathrm{i}$ is represented as a point in the complex plane with Cartesian coordinates (a, b). The horizontal axis corresponds to the real part of the complex number, while the vertical axis corresponds to the imaginary part. The line connecting the origin to the point z represents the vector form of the complex number.

Adding two complex numbers corresponds to vector addition. Given $z_1 = a + b\mathrm{i}$ and $z_2 = c + d\mathrm{i}$, their sum $z_1 + z_2 = (a + c) + (b + d)\mathrm{i}$ is the vector obtained by placing the tail of z_2 at the head of z_1 (or vice versa) and drawing the resultant vector from the origin to the new point. This is the same operation as vector addition in Euclidean geometry and can be visualised as the diagonal of a parallelogram formed by the two vectors.

Geometrically speaking, multiplying complex numbers combines two geometric effects: rotation and scaling. As outlined, if $z_1 = r_1 e^{\mathrm{i}\theta_1}$ and $z_2 = r_2 e^{\mathrm{i}\theta_2}$ in polar form, then their product is $r_1 r_2 e^{\mathrm{i}(\theta_1 + \theta_2)}$. This means the product has a modulus equal to the

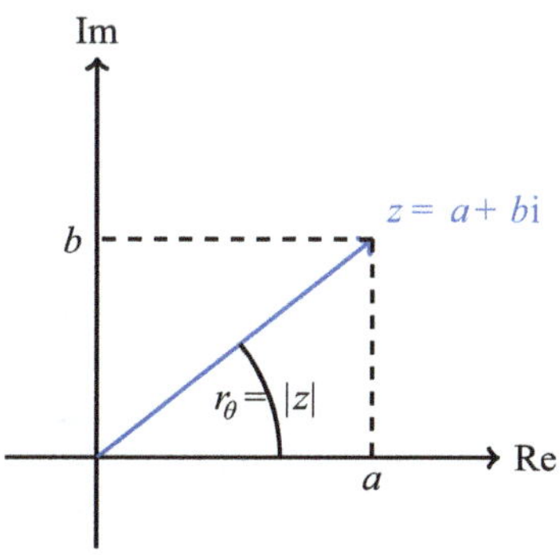

Figure 2.1. Argand diagram showing the complex number $z = a + b\mathrm{i}$.

product of the individual moduli, and an argument equal to the sum of the individual arguments. Geometrically, this corresponds to scaling the vector's length and rotating it counterclockwise by the sum of the two angles. This principle underlies many applications in oscillatory and signal systems, including those in cryocooler electronics and vibration analysis.

The vector has a length $r = |z|$, which is the modulus of the complex number, and it makes an angle $\theta = \arg(z)$ with the positive real axis, known as the argument of the complex number. The modulus is the same as the length of a vector from the origin to the point (a, b), making it useful in evaluating the magnitude of electrical signals, pressure amplitudes, or thermal responses in cryocooler components. The argument is measured in radians (or degrees) and represents the direction of the vector in the plane. The argument is essential in analysing phase relationships between periodic signals, a crucial aspect of cryocooler dynamics, where the phase angle between the mass flow and pressure pulse governs the thermodynamic cycle run by the cryocooler, as well as holding significance in control system timing and feedback circuitry. These two quantities; modulus and argument; allow the complex number to be expressed in polar form as

$$z = r(\cos\theta + \mathrm{i}\sin\theta).$$

The diagram also includes dashed lines projecting the point z onto the real and imaginary axes to highlight its components a and b. This visualisation reinforces the idea that polar and rectangular forms of complex numbers are simply two ways of describing the same point in the complex plane.

2.1.2.2 Advantages

The polar form of complex numbers offers several significant advantages, particularly in contexts where magnitude and angle are more insightful than Cartesian coordinates. One of its most powerful features is the simplification it brings to multiplication and division. When two complex numbers are expressed in polar form, their product is obtained by multiplying their moduli and adding their arguments. Similarly, their quotient is found by dividing the moduli and subtracting the arguments. This is far simpler and more intuitive than performing the same operations in rectangular form, which typically involves expanding and simplifying algebraic expressions.

Another key advantage of the polar form is its utility in raising complex numbers to powers and extracting roots. By applying de Moivre's theorem, which states that $(r\,\mathrm{e}^{\mathrm{i}\theta})^n = r^n\,\mathrm{e}^{\mathrm{i}n\theta}$, exponentiation becomes a straightforward matter of raising the modulus to the power and multiplying the angle. Finding roots involves taking the nth root of the modulus and dividing the argument appropriately, yielding multiple distinct but evenly spaced solutions in the complex plane.

Finally, the polar form is indispensable in modelling periodic and oscillatory phenomena. In disciplines exploited by cryocoolers such as electrical engineering, signal processing, and wave mechanics, complex exponentials are used to represent sinusoidal functions. The polar (or exponential) form captures both amplitude and phase information in a single, elegant expression. This makes it the natural language for analysing alternating current circuits, Fourier transforms, and wave propagation.

The geometric view of complex numbers using an Argand diagram transforms them from abstract algebraic objects into intuitive visual tools. This is particularly valuable in the context of cryocooler engineering, where periodic signals, control loops, and feedback dynamics often have natural geometric interpretations. Concepts such as phase margin, stability, and oscillation can be visualised and quantified using this geometric perspective, facilitating better understanding and design of cryogenic systems. This will be explored further in section 2.2.

Example
Consider the complex number $z = 1 + \mathrm{i}$. Here, $a = 1$ and $b = 1$:

$$\begin{aligned} r &= \sqrt{1^2 + 1^2} = \sqrt{2}, \\ \theta &= \tan^{-1}\left(\frac{1}{1}\right) = \frac{\pi}{4}. \end{aligned}$$

So the polar form is

$$z = \sqrt{2}\left(\cos\left(\frac{\pi}{4}\right) + \mathrm{i}\sin\left(\frac{\pi}{4}\right)\right).$$

In the next section, we will see how this form leads naturally into the exponential form $z = r\mathrm{e}^{\mathrm{i}\theta}$ using Euler's formula, which further simplifies many complex arithmetic operations.

2.1.3 Euler's formula and exponential form

One of the most elegant and profound relationships in mathematics is captured by Euler's formula, which connects exponential functions to trigonometric functions. Euler's formula is given by

$$\mathrm{e}^{\mathrm{i}\theta} = \cos\theta + \mathrm{i}\sin\theta. \tag{2.7}$$

This identity shows that a complex exponential can be expressed as a combination of sine and cosine functions, and it forms the cornerstone of the exponential form of complex numbers. It arises naturally when we expand both sides into their respective Taylor series and observe that the real and imaginary parts align perfectly.

By substituting Euler's formula into the polar representation of a complex number, we obtain the exponential form

$$z = r\mathrm{e}^{\mathrm{i}\theta}, \tag{2.8}$$

where $r = |z|$ is the modulus and $\theta = \arg(z)$ is the argument of the complex number. This compact representation encapsulates both the magnitude and direction of the complex number in a single expression. It is widely used across mathematics, physics, and engineering because of the simplification it brings to many types of calculations.

2.1.3.1 Multiplication and division

In exponential form, the multiplication and division of complex numbers become remarkably simple. Suppose we have two complex numbers,

$$z_1 = r_1 e^{i\theta_1},$$
$$z_2 = r_2 e^{i\theta_2}.$$

Then their product is

$$z_1 z_2 = r_1 r_2 e^{i(\theta_1 + \theta_2)}, \tag{2.9}$$

and their quotient is

$$\frac{z_1}{z_2} = \frac{r_1}{r_2} e^{i(\theta_1 - \theta_2)}. \tag{2.10}$$

This clearly illustrates the geometric interpretation of complex multiplication as a combination of scaling (multiplying moduli) and rotation (adding arguments).

2.1.3.2 Exponentiation and roots

The exponential form is especially powerful when raising complex numbers to integer or fractional powers. Using de Moivre's theorem, we can write

$$z^n = (re^{i\theta})^n = r^n e^{in\theta}. \tag{2.11}$$

Similarly, taking the nth root of a complex number yields

$$z_k = r^{1/n} e^{i(\theta + 2k\pi)/n}, \quad k = 0, 1, \ldots, n - 1, \tag{2.12}$$

revealing the n distinct roots spaced evenly around a circle in the complex plane.

2.1.3.3 Applications

Euler's formula and the exponential form of complex numbers are foundational tools in the analysis and modelling of cryogenic systems, particularly in cryocooler design and control. In the electrical subsystems of cryocoolers, such as drive electronics for compressors or active magnetic regenerator circuits, alternating current signals are frequently represented using complex exponentials to simplify the analysis of phase shifts, impedance, and resonant behaviour. The phasor representation afforded by Euler's formula is especially helpful when studying sinusoidal steady-state conditions, allowing engineers to manage time-dependent variables with algebraic clarity.

Moreover, in the control and regulation of cryocoolers, such as pulse tube or Stirling-cycle systems, the exponential form of complex numbers plays a vital role in the application of Laplace and Fourier transforms. These transforms enable the modelling of dynamic thermal and mechanical responses to perturbations, allowing for system identification, stability analysis, and frequency-domain design of feedback controllers. Oscillatory behaviours within regenerator flow fields and pressure waves in working gas columns can also be conveniently represented using expressions like $e^{i\omega t}$.

In addition, numerical simulations of cryocooler dynamics often require solving linear time-invariant differential equations, where complex eigenvalues naturally arise and are best interpreted using exponential forms. Whether modelling phase lags in thermodynamic cycles or analysing the frequency response of a thermal transfer function, Euler's identity provides an elegant and practical mathematical tool.

Ultimately, Euler's formula is more than just a mathematical identity: it is a profound connection between seemingly unrelated functions that reveals the underlying unity of mathematics. In the context of cryocoolers, it bridges the theoretical and the applied, enabling a deeper understanding of the physical processes governing low-temperature refrigeration systems.

2.1.4 Powers and roots of complex numbers

The exponential and polar forms of complex numbers make it especially straightforward to compute powers and roots, operations that are otherwise tedious in rectangular (Cartesian) form. These operations are grounded in two key results: de Moivre's theorem for powers and a general formula for extracting roots.

2.1.4.1 de Moivre's theorem

de Moivre's theorem provides a powerful method for raising complex numbers to integer powers. It states that for any integer n, the following identity holds:

$$(\cos\theta + \mathrm{i}\sin\theta)^n = \cos(n\theta) + \mathrm{i}\sin(n\theta). \tag{2.13}$$

This can be derived using mathematical induction or directly from Euler's formula:

$$z = r\mathrm{e}^{\mathrm{i}\theta} \quad \Rightarrow \quad z^n = (r\mathrm{e}^{\mathrm{i}\theta})^n = r^n\mathrm{e}^{\mathrm{i}n\theta}. \tag{2.14}$$

This elegant result tells us that raising a complex number to the power n simply requires raising the modulus to n and multiplying the argument by n. This interpretation is particularly useful when analysing periodic signals, oscillatory systems, and resonant behaviour in cryocooler electronics.

As an example, consider $z = 2\mathrm{e}^{\mathrm{i}\frac{\pi}{4}}$. Then

$$z^3 = 2^3\mathrm{e}^{\mathrm{i}3\cdot\frac{\pi}{4}} = 8\mathrm{e}^{\mathrm{i}\frac{3\pi}{4}}.$$

This corresponds to a vector of length 8 at an angle 135° in the complex plane.

2.1.4.2 Roots of complex numbers

Finding the nth roots of a complex number is equally elegant in exponential form. Let $z = r\mathrm{e}^{\mathrm{i}\theta}$ be a complex number. Then the n distinct nth roots of z are given by

$$z_k = r^{1/n}\mathrm{e}^{\mathrm{i}(\theta+2k\pi)/n}, \quad k = 0, 1,\ldots,n-1. \tag{2.15}$$

This formula shows that the nth roots of a complex number are evenly spaced around a circle of radius $r^{1/n}$ in the complex plane. The angle between successive roots is $\frac{2\pi}{n}$ radians. These roots form the vertices of a regular n-gon centred at the origin.

As an example, let us find the cube roots of $z = 8e^{i0}$. The modulus is $r = 8$, and the argument is $\theta = 0$. The cube roots are

$$\begin{aligned} z_0 &= 2e^{i0} = 2, \\ z_1 &= 2e^{i\frac{2\pi}{3}} = 2\left(\cos\frac{2\pi}{3} + i\sin\frac{2\pi}{3}\right), \\ z_2 &= 2e^{i\frac{4\pi}{3}} = 2\left(\cos\frac{4\pi}{3} + i\sin\frac{4\pi}{3}\right). \end{aligned}$$

These three roots are located at 120° intervals on the circle of radius 2 in the complex plane.

In the context of cryocoolers, such root-finding techniques are valuable in the analysis of characteristic equations arising in dynamic thermal and mechanical models. For example, roots of characteristic polynomials determine the natural frequencies and stability of oscillatory components such as displacer pistons, pulse tubes, and compressor chambers. Being able to compute and interpret complex roots allows engineers to assess phase relationships, damping behaviour, and mode interactions in frequency-domain analyses.

In summary, de Moivre's theorem and the root formula highlight the power and elegance of the exponential form, offering not only mathematical simplicity but also deep insight into the geometry and behaviour of complex systems such as oscillating flow cryocoolers.

2.1.5 Applications of complex numbers in cryocoolers

Complex numbers are deeply embedded in the mathematical modelling and analysis of cryocooler systems. One of the most direct applications arises in electrical engineering, where complex numbers are used to represent alternating current (AC) signals. In cryocoolers, AC signals frequently occur in the drive electronics of Stirling and pulse tube coolers. By expressing these sinusoidal voltages and currents as complex exponentials, engineers can analyse phase relationships, impedance, and power transfer with greater clarity and simplicity.

In control systems and signal processing, which are integral to the regulation and optimisation of cryocooler performance, complex numbers provide a framework for analysing system stability and response characteristics. Techniques such as the Laplace and Fourier transforms leverage complex exponentials to convert time-domain behaviour into the frequency domain. This is especially useful in identifying resonant frequencies, damping ratios, and designing compensators or filters in feedback loops.

Furthermore, complex numbers also find application in fluid dynamics and quantum mechanics, both of which intersect with cryogenic technologies. In fluid dynamics, especially within regenerators and oscillatory gas columns, pressure and velocity profiles can be modelled using complex-valued functions to describe harmonic motion and standing waves. Similarly, in the quantum mechanical treatment of low-temperature phenomena, such as those relevant to cryogenic sensors or superconducting devices, the wave function is inherently complex and governed by equations such as Schrödinger's, where solutions involve exponential and sinusoidal components.

A central mathematical tool that underpins many of these applications is the concept of phasors, which represent sinusoidal signals as rotating vectors in the complex plane. Phasors greatly simplify calculations involving phase shifts and amplitude modulation and are ubiquitous in the analysis of harmonic systems in cryocooler engineering. This concept will be explored in greater detail in section 2.2.

2.2 Phasors

Phasors are a fundamental mathematical tool for representing and analysing sinusoidal signals using complex numbers. They are especially valuable in the context of cryocooler systems, where alternating current (AC) signals, oscillatory mechanical motions, and dynamic thermal responses often exhibit harmonic behaviour. By representing time-varying functions as rotating vectors in the complex plane, phasors enable a transition from the time domain to the frequency domain, where complex algebra replaces differential calculus.

2.2.1 From sinusoids to phasors

A sinusoidal signal of the form

$$x(t) = A\cos(\omega t + \phi) \tag{2.16}$$

can be re-expressed using Euler's identity as the real part of a complex exponential:

$$x(t) = \Re\{A\mathrm{e}^{\mathrm{i}(\omega t+\phi)}\} = \Re\left\{\underbrace{A\mathrm{e}^{\mathrm{i}\phi}}_{\text{phasor}} \cdot \mathrm{e}^{\mathrm{i}\omega t}\right\}. \tag{2.17}$$

The complex constant $A\mathrm{e}^{\mathrm{i}\phi}$ is called the phasor representation of the signal, encoding both amplitude and phase shift. This separation allows one to handle sinusoidal phenomena algebraically in the frequency domain, greatly simplifying analysis in steady-state conditions.

2.2.2 Phasor notation and manipulation

Phasor notation, also known as angle notation, is a concise mathematical tool widely used in electronics and signal analysis to represent sinusoidal signals. It expresses a sinusoid by separating its amplitude and phase from its time-varying behaviour, thereby simplifying many mathematical operations.

A phasor is defined by its magnitude A and phase angle θ, and is typically written as

$$A\angle\theta, \tag{2.18}$$

where the angle θ is assumed to be in radians unless explicitly stated otherwise. For instance, the phasor $1\angle 90^{\circ}$ describes a unit-magnitude vector oriented at 90° in the complex plane, which corresponds to the complex number

$$\cos\theta + \mathrm{i}\sin\theta = \mathrm{e}^{\mathrm{i}\theta}, \tag{2.19}$$

via Euler's identity. This representation geometrically places the number on the unit circle at angle θ from the real axis.

For example:

$$1\angle 90^\circ = (0,\ 1) = \mathrm{i}. \tag{2.20}$$

The general time-dependent real-valued sinusoid is

$$A\cos(\omega t + \theta), \tag{2.21}$$

where A is the amplitude, ω is the angular frequency, and θ is the phase offset. Here, only t varies with time. To represent this sinusoid using phasor methods, we extend it to the complex domain by including an imaginary component

$$\mathrm{i}A\sin(\omega t + \theta), \tag{2.22}$$

so that the complex expression becomes

$$A\cos(\omega t + \theta) + \mathrm{i}A\sin(\omega t + \theta) = A\mathrm{e}^{\mathrm{i}(\omega t + \theta)}. \tag{2.23}$$

This separates the sinusoid into a rotating exponential $\mathrm{e}^{\mathrm{i}\omega t}$ and a static complex amplitude $A\mathrm{e}^{\mathrm{i}\theta}$. The full sinusoid can then be represented as

$$\Re\{A\mathrm{e}^{\mathrm{i}(\omega t + \theta)}\} = A\cos(\omega t + \theta), \tag{2.24}$$

which simply takes the real part of the complex exponential to recover the original signal.

This approach enables algebraic manipulations to be performed entirely on the phasor $A\mathrm{e}^{\mathrm{i}\theta}$, greatly simplifying calculations in linear systems. The time-dependent term $\mathrm{e}^{\mathrm{i}\omega t}$ is factored out during intermediate steps and reintroduced at the end to obtain the physical solution.

Thus, the function $A\mathrm{e}^{\mathrm{i}(\omega t + \theta)}$ acts as a compact analytic form of the sinusoid $A\cos(\omega t + \theta)$, and it is often convenient to refer to this function as a phasor in engineering applications.

2.2.2.1 Multiplication and division in phasor notation

Operations such as multiplying and dividing phasors become straightforward when using polar representation. If we have two phasors

$$\vec{v}_1 = A_1\angle\theta_1, \tag{2.25}$$

$$\vec{v}_2 = A_2\angle\theta_2, \tag{2.26}$$

then their product is

$$\vec{v}_1 \cdot \vec{v}_2 = A_1 A_2\angle(\theta_1 + \theta_2), \tag{2.27}$$

and their quotient is

$$\frac{\vec{v}_1}{\vec{v}_2} = \frac{A_1}{A_2}\angle(\theta_1 - \theta_2). \tag{2.28}$$

These operations reflect the rules of complex number arithmetic in polar form: multiplication results in multiplying the magnitudes and adding the angles; division results in dividing the magnitudes and subtracting the angles.

2.2.2.2 Multiplication by a complex constant

Consider a phasor represented in exponential form as $Ae^{i\theta}e^{i\omega t}$, where A is the amplitude, θ is the phase angle, and ω is the angular frequency. When this phasor is multiplied by another complex constant $Be^{i\phi}$, the result is another phasor, differing only in amplitude and phase:

$$\mathrm{Re}[(Ae^{i\theta} \cdot Be^{i\phi})e^{i\omega t}] = \mathrm{Re}[(ABe^{i(\theta+\phi)})e^{i\omega t}] \tag{2.29}$$

$$= AB\cos(\omega t + (\theta + \phi)). \tag{2.30}$$

This operation is common in electrical engineering, where $Be^{i\phi}$ may represent a time-invariant complex impedance. Multiplying a current phasor by such an impedance yields the voltage phasor. However, phasor multiplication must be used carefully, as squaring a phasor or multiplying two phasors corresponds to combining sinusoids at the same frequency; a nonlinear operation if interpreted in the time domain.

2.2.2.3 Addition of phasors

Adding two phasors of equal frequency results in another sinusoidal function of that frequency. Let the two input waveforms be

$$A_1\cos(\omega t + \theta_1) + A_2\cos(\omega t + \theta_2). \tag{2.31}$$

This sum can be interpreted using Euler's identity and the real part of exponentials:

$$= \mathrm{Re}[A_1e^{i\theta_1}e^{i\omega t} + A_2e^{i\theta_2}e^{i\omega t}] \tag{2.32}$$

$$= \mathrm{Re}[(A_1e^{i\theta_1} + A_2e^{i\theta_2})e^{i\omega t}] \tag{2.33}$$

$$= \mathrm{Re}[A_3e^{i\theta_3}e^{i\omega t}] = A_3\cos(\omega t + \theta_3), \tag{2.34}$$

where $A_3e^{i\theta_3} = A_1e^{i\theta_1} + A_2e^{i\theta_2}$. The resultant phasor has a magnitude and phase defined by

$$A_3^2 = (A_1\cos\theta_1 + A_2\cos\theta_2)^2 + (A_1\sin\theta_1 + A_2\sin\theta_2)^2, \tag{2.35}$$

and angle

$$\theta_3 = \begin{cases} \arctan\left(\dfrac{A_1 \sin\theta_1 + A_2 \sin\theta_2}{A_1 \cos\theta_1 + A_2 \cos\theta_2}\right), & \text{if} A_1 \cos\theta_1 + A_2 \cos\theta_2 > 0 \\ \pi + \arctan\left(\dfrac{A_1 \sin\theta_1 + A_2 \sin\theta_2}{A_1 \cos\theta_1 + A_2 \cos\theta_2}\right), & \text{if} A_1 \cos\theta_1 + A_2 \cos\theta_2 < 0 \cdot \\ \text{sgn}(A_1 \sin\theta_1 + A_2 \sin\theta_2) \cdot \dfrac{\pi}{2}, & \text{if} A_1 \cos\theta_1 + A_2 \cos\theta_2 = 0 \end{cases} \tag{2.36}$$

Alternatively, using the law of cosines for phasors:

$$A_3^2 = A_1^2 + A_2^2 + 2A_1A_2 \cos(\theta_1 - \theta_2). \tag{2.37}$$

This is written succinctly in angle notation as

$$A_1\angle\theta_1 + A_2\angle\theta_2 = A_3\angle\theta_3. \tag{2.38}$$

This result can be intuitively interpreted using a geometric diagram of phasor addition in the complex plane. Figure 2.2 illustrates this process, where each rotating vector (phasor) represents a sinusoidal function of the same frequency but with distinct amplitudes and phases. Individual phasors are shown in green and blue, and their resultant, formed by vector addition, is shown in red.

In the lower portion of the figure, the vectors are visualised in the complex plane with their corresponding amplitudes and angular displacements. The projections of each phasor onto the real axis yield the instantaneous values of the original cosine functions, while their vector sum determines the net waveform. The upper portion of the diagram shows how these individual signals combine in the time domain: the green and blue curves represent the time-varying signals $z_1(t)$ and $z_2(t)$, and the red curve corresponds to the resulting signal $z(t)$ obtained from their superposition.

This dual visualisation bridges the connection between time-domain waveform addition and complex-plane vector addition, highlighting the utility of phasors for interpreting interference, phase relationships, and superposition in oscillatory systems.

This principle of phasor addition explains how sinusoids combine in physical systems. When two or more sinusoids of the same frequency are superimposed, the resulting waveform is also a sinusoid of that frequency, with an amplitude and phase determined by the vector addition of the corresponding phasors.

This behaviour underpins phenomena like constructive and destructive interference. For instance, if three sinusoids are offset in phase by 120°, their sum is zero:

$$\cos(\omega t) + \cos\left(\omega t + \frac{2\pi}{3}\right) + \cos\left(\omega t - \frac{2\pi}{3}\right) = 0. \tag{2.39}$$

This case occurs in three-phase power systems and in diffraction patterns.

In rotating vector representations, a phasor completes a full revolution of 2π radians per cycle. The tip of the phasor traces a sinusoid when plotted against time,

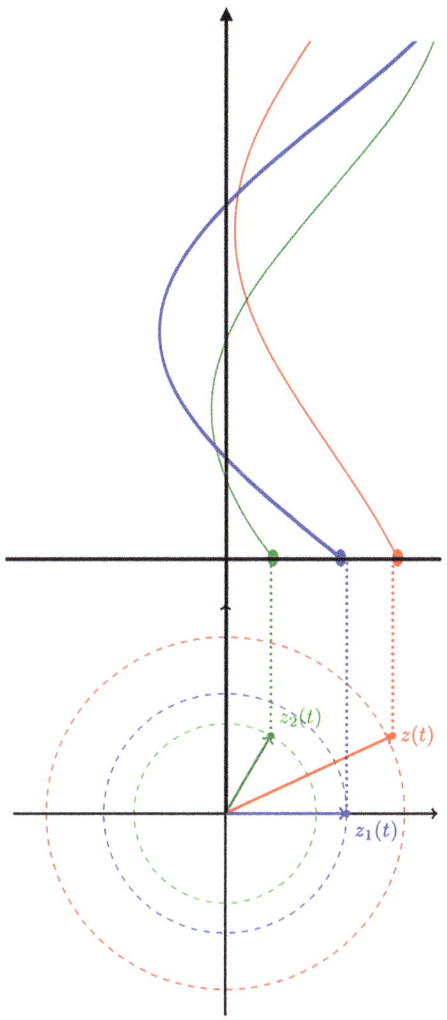

Figure 2.2. Phasor diagram and corresponding time-domain waveforms showing the addition of two sinusoidal signals $z_1(t)$ and $z_2(t)$.

with the horizontal axis representing time and the vertical axis representing the instantaneous value. The angle θ at any time t defines the position of the phasor.

For instance, when the phasor is at 0°, it is on the real axis; at 90°, it reaches the positive imaginary peak; at 180°, it returns to the negative real axis, and at 270°, it reaches the negative imaginary peak.

When analysing alternating signals such as voltage or current, the phasor position at $t = 0$ is often taken as the reference. However, if the signal starts with a delay or leads the reference, this introduces a phase shift Φ, which must be represented explicitly in the phasor model to compare multiple waveforms accurately.

2.2.3 Phasor differentiation and integration

In phasor analysis, both differentiation and integration operations on sinusoidal functions can be elegantly handled by algebraic manipulation, due to the exponential form of complex sinusoids. Specifically, taking the derivative or integral of a phasor yields another phasor, with predictable modifications to its amplitude and phase.

2.2.3.1 Differentiation of a phasor

Let us consider a time-varying complex sinusoid represented as

$$\mathfrak{R}(Ae^{i\theta} \cdot e^{i\omega t}).$$

To compute its time derivative, we apply the product rule:

$$\frac{d}{dt}\mathfrak{R}(Ae^{i\theta} \cdot e^{i\omega t}) = \mathfrak{R}\left(Ae^{i\theta} \cdot \frac{d}{dt}e^{i\omega t}\right) \tag{2.40}$$

$$=\mathfrak{R}(Ae^{i\theta} \cdot i\omega e^{i\omega t}) \tag{2.41}$$

$$=\mathfrak{R}(i\omega Ae^{i(\theta+\omega t)}) \tag{2.42}$$

$$=\omega A\cos\left(\omega t + \theta + \frac{\pi}{2}\right). \tag{2.43}$$

This shows that differentiation introduces a phase shift of $\frac{\pi}{2}$ and scales the amplitude by ω. Thus, differentiating a sinusoid corresponds to multiplying its phasor by $i\omega$.

2.2.3.2 Integration of a phasor

Analogously, integration introduces a phase shift of $-\frac{\pi}{2}$, and scales the amplitude by $\frac{1}{\omega}$. That is, integrating a phasor in time is equivalent to

$$\frac{1}{i\omega} = \frac{e^{-i\pi/2}}{\omega}.$$

2.2.4 Differential equations and phasors

One of the most powerful aspects of phasor analysis lies in its ability to convert differential operations into straightforward algebraic manipulations. This transformation is particularly valuable in engineering disciplines where systems are governed by ordinary differential equations (ODEs) involving sinusoidal inputs, such as thermal and mechanical oscillators in cryogenic environments.

Suppose we consider a real-valued sinusoidal function $x(t)$ that varies harmonically with time. If the phasor representation of $x(t)$ is given by the complex quantity $\tilde{X}$, then the time-domain function can be expressed as $x(t) = \mathfrak{R}\{\widetilde{X}e^{i\omega t}\}$, where ω is the angular frequency of the oscillation. The act of differentiating this function with respect to time results in the following identity:

$$\frac{\mathrm{d}}{\mathrm{d}t}x(t) = \Re\{\mathrm{i}\omega\tilde{X}\mathrm{e}^{\mathrm{i}\omega t}\}. \tag{2.44}$$

This demonstrates that taking the first derivative of a sinusoidal signal is equivalent to multiplying its phasor by $\mathrm{i}\omega$, while maintaining the same exponential time dependence. The operation becomes purely algebraic in the phasor domain.

Extending this idea further, the second derivative of the function $x(t)$ becomes

$$\frac{\mathrm{d}^2}{\mathrm{d}t^2}x(t) = \Re\{-\omega^2\tilde{X}\mathrm{e}^{\mathrm{i}\omega t}\}. \tag{2.45}$$

Here, the phasor $\tilde{X}$ is now multiplied by $-\omega^2$, and again, this allows a second-order differential operation to be translated into a simple scalar multiplication in the complex domain. As a result, even higher-order derivatives follow a predictable algebraic pattern in the phasor framework.

This algebraic substitution greatly simplifies the analysis and solution of linear time-invariant (LTI) systems. In cryogenic engineering, for instance, temperature fluctuations in sensitive components such as cold fingers or focal plane arrays are often driven by periodic disturbances, whether from active heater modulation, cryocooler harmonics, or external thermal loads. Instead of solving time-domain differential equations directly, which may be analytically complex or numerically intensive, engineers can apply phasor techniques to model these periodic behaviours using algebraic expressions that are easier to manipulate and solve.

By reducing differential equations to algebraic forms, phasors provide an elegant and computationally efficient means of characterising the dynamic response of cryogenic systems. This includes not only thermal oscillations, but also mechanical resonances, vibration modes, and active control loops—all of which may be subject to sinusoidal forcing and therefore lend themselves naturally to analysis in the phasor domain.

2.2.5 Electrical impedance in the phasor domain

In alternating current (AC) circuit analysis, phasor methods offer a powerful way to extend Ohm's law into the frequency domain. Rather than dealing with time-varying voltages and currents directly, phasors represent these sinusoidal signals as complex numbers, greatly simplifying the analysis of linear electrical systems. In the phasor domain, Ohm's law is written as

$$\tilde{V} = \tilde{I}Z, \tag{2.46}$$

where $\tilde{V}$ and $\tilde{I}$ denote the phasor representations of the voltage and current, respectively, and Z is the complex impedance of the circuit element. This form highlights that in the frequency domain, voltage and current remain linearly related, but through a frequency-dependent impedance rather than a simple resistance.

Each passive component in an electrical circuit contributes a specific form of impedance. A pure resistor offers no frequency dependence and simply contributes a real impedance value,

$$Z_R = R, \tag{2.47}$$

where R is the resistance in ohms. This means that the current and voltage remain in phase when passing through a resistor, and no reactive effects occur.

An inductor, in contrast, introduces a reactive component that is proportional to the rate of change of current. In the phasor domain, this is captured by the impedance expression

$$Z_L = \mathrm{i}\omega L, \tag{2.48}$$

where L is the inductance and ω is the angular frequency of the AC signal. The imaginary unit i denotes that the voltage across an inductor leads the current by 90 degrees, or one-quarter of a cycle. As frequency increases, the impedance of the inductor increases linearly, effectively resisting rapid changes in current.

A capacitor, on the other hand, behaves inversely with respect to frequency. Its impedance in the phasor domain is given by

$$Z_C = \frac{1}{\mathrm{i}\omega C}, \tag{2.49}$$

where C is the capacitance. This expression indicates that capacitors oppose low-frequency signals more than high-frequency ones, allowing rapidly changing signals to pass more easily. In terms of phase, the current through a capacitor leads the voltage by 90 degrees, again illustrated by the imaginary nature of its impedance.

These impedance relationships are particularly relevant in the context of cryocooler driver circuits, which often employ oscillatory signals to drive compressor motors or pulse tube actuators at specific resonant frequencies. The ability to model and manipulate circuit behaviour in the phasor domain enables engineers to design frequency-selective networks that can match impedance for efficient power delivery, filter out unwanted harmonics, and prevent signal distortion or overheating. Because cryocoolers often operate within narrow frequency bands for optimal thermodynamic performance, understanding impedance behaviour in the phasor domain is essential for ensuring that electrical energy is transferred with minimal loss and maximum stability across these crucial interfaces.

2.2.6 Mechanical analogs and cryocooler dynamics

Phasor analysis is not confined to electrical systems; it extends naturally to mechanical systems, especially those governed by harmonic motion. In the context of cryocooler engineering, where components such as displacers, pistons, and compressors exhibit oscillatory behaviour, phasors provide a powerful means of representing and analysing mechanical dynamics. A common starting point is the harmonic displacement of a mechanical component, expressed in the time domain as $x(t) = X\cos(\omega t + \phi)$, where X is the amplitude, ω is the angular frequency, and ϕ is the phase offset. The phasor representation of this displacement is given by $\widetilde{X} = X\mathrm{e}^{\mathrm{i}\phi}$, compactly encoding both magnitude and phase information in a single complex quantity.

Once the displacement is represented in phasor form, its derivatives with respect to time, such as velocity and acceleration, can be computed using simple algebraic operations. The velocity, being the first derivative of displacement, becomes

$$\tilde{v} = \mathrm{i}\omega\tilde{X}, \tag{2.50}$$

while the acceleration, as the second derivative, takes the form

$$\tilde{a} = -\omega^2\tilde{X}. \tag{2.51}$$

These expressions illustrate one of the main benefits of phasor analysis: differential operations that would otherwise involve time-varying functions become straightforward multiplications by constants in the frequency domain.

Applying Newton's second law, $F = Ma$, within the phasor framework, and incorporating additional mechanical effects such as damping and stiffness, yields a frequency-domain expression for the total force acting on the system. Specifically, the total force phasor is given by

$$\tilde{F} = (K - \omega^2 M + \mathrm{i}\omega C)\tilde{X}, \tag{2.52}$$

where K is the spring constant representing stiffness, M is the mass of the oscillating element, and C is the damping coefficient. This formulation is a direct mechanical analog of electrical impedance, with each term representing a mechanical component's frequency-dependent opposition to motion: stiffness K corresponds to a restoring force, inertia M opposes acceleration, and damping C resists velocity.

In cryocooler systems, this model is invaluable for analysing the behaviour of internal moving parts such as linear compressors or pulse tube displacers. These components operate at precise frequencies to maximise thermodynamic efficiency, and any deviation from the intended frequency response, due to resonance, excess damping, or mechanical nonlinearity, can result in reduced cooling performance or mechanical instability. By modelling such elements as mass–spring–damper systems and applying phasor methods, engineers can predict the amplitude and phase of displacements and forces throughout the cryocooler cycle. This enables the design of more robust, quieter, and longer-lasting cryogenic systems, particularly for sensitive applications such as space telescopes or infrared detectors, where vibration and thermal drift must be rigorously controlled.

Moreover, the phasor framework facilitates the integration of mechanical models with their electrical analogs, supporting holistic multiphysics simulations of cryocooler performance. This unified perspective is essential for designing cryocoolers that respond predictably to both electrical inputs and mechanical loads, ensuring that performance targets are achieved across a range of environmental and operational conditions.

2.2.7 Utility in cryogenic systems

Phasors are indispensable in the mathematical modelling of cryocoolers, particularly in analysing drive electronics and oscillatory mechanical subsystems. Components such as compressors, pistons, and regenerators often experience periodic excitations that can be accurately represented by sinusoidal signals. Using phasors, these oscillations can be described using complex numbers, facilitating the analysis of amplitude, phase, and frequency responses.

In electrical circuits, phasors describe voltage and current waveforms, and are instrumental in designing cryocooler driver circuits. In mechanical domains, phasors simplify the modelling of vibrational behaviour and resonance effects in reciprocating structures.

2.2.8 Application: first-order phasor response in cryocooler heater control

In cryocooler systems, particularly in space-based Stirling or pulse tube coolers, precise thermal regulation is critical to achieving stable operation at cryogenic temperatures. Many cryogenic stages incorporate electrical heaters as part of an active temperature control loop. When these heaters are driven with sinusoidal or modulated input power, the resulting temperature response can be modelled as a first-order thermal system. This behaviour is mathematically analogous to the response of a resistor–capacitor (RC) circuit in electronics.

2.2.8.1 Thermal analogy to the RC circuit

Let $T(t)$ represent the time-varying temperature of the cold tip and $Q_{\text{in}}(t)$ the input power applied by the heater. The thermal system's dynamics can be described using a first-order linear differential equation:

$$\frac{\mathrm{d}T(t)}{\mathrm{d}t} + \frac{1}{RC}T(t) = \frac{1}{RC}Q_{\text{in}}(t), \tag{2.53}$$

where R is the thermal resistance (in K W $^{-1}$), C is the thermal capacitance or heat capacity (in J K^{-1}), and their product RC defines the characteristic thermal time constant (in s). This formulation mirrors that of an RC electrical circuit, with temperature analogous to voltage and heat input analogous to current.

2.2.8.2 Sinusoidal heater input and phasor transformation

Assuming the heater is modulated with a sinusoidal power input,

$$Q_{\text{in}}(t) = Q_0 \cos(\omega t + \theta), \tag{2.54}$$

we convert this to phasor form using the exponential representation:

$$\tilde{Q}_{\text{in}} = Q_0 \mathrm{e}^{\mathrm{i}\theta}. \tag{2.55}$$

We seek a steady-state solution for the temperature of the form

$$T(t) = \mathrm{Re}\left\{\tilde{T}\mathrm{e}^{\mathrm{i}\omega t}\right\}, \tag{2.56}$$

where $\tilde{T}$ is the phasor representation of the temperature.

Substituting this into the differential equation and dividing out the common exponential term $\mathrm{e}^{\mathrm{i}\omega t}$ leads to

$$\mathrm{i}\omega\tilde{T} + \frac{1}{RC}\tilde{T} = \frac{1}{RC}\tilde{Q}_{\text{in}}. \tag{2.57}$$

Solving for the phasor temperature response $\tilde{T}$ gives

$$\tilde{T} = \frac{1}{\mathrm{i}\omega + \frac{1}{RC}} \cdot \tilde{Q}_{\mathrm{in}} = \frac{R}{1 + \mathrm{i}\omega RC} \cdot Q_0 \mathrm{e}^{\mathrm{i}\theta}. \tag{2.58}$$

2.2.8.3 Amplitude and phase response

To understand the effect of the system on both the amplitude and phase of the output temperature, we express the frequency response factor in polar form:

$$\frac{1}{1 + \mathrm{i}\omega RC} = \frac{1}{\sqrt{1 + (\omega RC)^2}} \mathrm{e}^{-\mathrm{i}\phi(\omega)}, \tag{2.59}$$

where the phase lag $\phi(\omega)$ is given by

$$\phi(\omega) = \tan^{-1}(\omega RC). \tag{2.60}$$

The final steady-state temperature response is therefore

$$T(t) = \frac{RQ_0}{\sqrt{1 + (\omega RC)^2}} \cos(\omega t + \phi - \arctan(\omega\ R\ C)). \tag{2.61}$$

2.2.8.4 Physical interpretation

The most significant feature of this expression is the phase lag $\phi(\omega)$ between the heater input and the resulting temperature. This lag represents the time delay introduced by the finite thermal inertia of the system. As the frequency of the input increases, the phase lag grows, approaching 90° for very high frequencies. This reflects the inability of the cold tip to respond instantaneously to rapid changes in heater power, which is a critical consideration in high-precision control systems.

Additionally, the amplitude of the temperature response is attenuated relative to the input power by a factor of $1/\sqrt{1 + (\omega RC)^2}$. This attenuation increases with frequency, confirming that the system behaves as a low-pass filter. Low-frequency components pass through with little distortion, while higher frequencies are suppressed. This behaviour is particularly beneficial in cryocooler systems, where high-frequency thermal oscillations, often arising from mechanical vibrations or PWM heater drives, can induce instability or degrade detector performance.

The thermal time constant RC thus acts as a tuning parameter: a larger RC smooths out fluctuations more effectively but also slows the system's response time. This tradeoff must be balanced carefully in mission design, particularly when fast thermal transitions or feedback loops are required.

Overall, phasor analysis provides a compact and insightful means of predicting and optimising thermal responses in cryocooler heater systems. By treating the thermal system as a first-order filter, we gain quantitative understanding of amplitude suppression, phase lag, and frequency-dependent behaviour, all of which are central to achieving stable and responsive cryogenic operation.

2.2.9 Worked example: resonance in LC cryocooler circuit

Consider a simple LC series circuit driving a cryocooler actuator. The total impedance is

$$Z = \mathrm{i}\omega L + \frac{1}{\mathrm{i}\omega C}. \tag{2.62}$$

At the resonant frequency

$$\omega_0 = \frac{1}{\sqrt{LC}}, \tag{2.63}$$

impedance becomes purely resistive. In phasor terms, voltage and current align in phase, indicating resonance. At this point, energy oscillates efficiently between magnetic and electric fields, with minimal energy dissipation, a critical condition for maximising mechanical actuation in cryocooler components.

2.3 Expansion series

Many of the physical phenomena encountered in cryocooler modelling, such as temperature distributions, oscillating pressures, and transient thermal loads, are inherently governed by functions that may not be expressible in closed form. In these cases, *expansion series* serve as an indispensable tool, allowing us to approximate complex functions with polynomials or orthogonal basis sets to arbitrary precision over well-defined domains. This section outlines the primary expansion methods relevant to cryocooler analysis, focusing on Taylor series, Maclaurin series, and generalised orthogonal expansions such as Fourier and Bessel series.

2.3.1 Taylor and Maclaurin series

The Taylor series provides a local polynomial approximation of a smooth function about a point $x = a$. For a function $f(x)$ that is infinitely differentiable near a, we write

$$f(x) = \sum_{n=0}^{\infty} \frac{f^{(n)}(a)}{n!}(x - a)^n. \tag{2.64}$$

In the special case where $a = 0$, this becomes the Maclaurin series:

$$f(x) = \sum_{n=0}^{\infty} \frac{f^{(n)}(0)}{n!}x^n. \tag{2.65}$$

These expansions are particularly useful for approximating thermodynamic state equations or modelling temperature-dependent material properties, such as thermal conductivity or heat capacity, over small ranges of variation.

Example: The exponential decay term commonly found in thermal response models,

$$\mathrm{e}^{-\alpha t},$$

has the Maclaurin expansion

$$e^{-\alpha t} = \sum_{n=0}^{\infty} \frac{(-\alpha t)^n}{n!}.$$

This becomes useful when solving differential equations using power series methods, especially in lumped-parameter or low-order analytical models.

2.3.2 Orthogonal function expansions

In spatially extended systems or in time-harmonic steady-state problems, it is often more efficient to expand the target function in terms of an orthogonal basis over a finite interval. A general orthogonal expansion takes the form

$$f(x) = \sum_{n=1}^{\infty} a_n \phi_n(x), \tag{2.66}$$

where $\{\phi_n(x)\}$ is a complete orthogonal set with respect to some inner product defined over a domain $[a, b]$. The coefficients a_n are computed via

$$a_n = \frac{\langle f, \phi_n \rangle}{\langle \phi_n, \phi_n \rangle}. \tag{2.67}$$

Two common types of orthogonal expansions used in cryocooler modelling include Fourier and Bessel series, discussed below.

2.3.2.1 Fourier series

For periodic functions $f(x)$ defined on $[-L, L]$, the Fourier series expansion is

$$f(x) = a_0 + \sum_{n=1}^{\infty} \left[a_n \cos\left(\frac{n\pi x}{L}\right) + b_n \sin\left(\frac{n\pi x}{L}\right) \right], \tag{2.68}$$

where the coefficients are given by

$$a_0 = \frac{1}{2L} \int_{-L}^{L} f(x) \mathrm{d}x, \tag{2.69}$$

$$a_n = \frac{1}{L} \int_{-L}^{L} f(x) \cos\left(\frac{n\pi x}{L}\right) \mathrm{d}x, \tag{2.70}$$

$$b_n = \frac{1}{L} \int_{-L}^{L} f(x) \sin\left(\frac{n\pi x}{L}\right) \mathrm{d}x. \tag{2.71}$$

Fourier expansions are indispensable in transient and steady-state heat transfer analyses involving oscillatory or cyclic boundary conditions, such as the periodic compression and expansion cycles in Stirling and pulse tube cryocoolers.

2.3.2.2 Bessel series

When the system geometry exhibits cylindrical symmetry, such as in coaxial regenerators or annular gas columns, solutions to the governing PDEs often involve Bessel functions. The Bessel series expansion for a function $f(r)$ defined over $0 \leqslant r \leqslant R$ is

$$f(r) = \sum_{n=1}^{\infty} A_n J_0\left(\frac{\alpha_n r}{R}\right), \tag{2.72}$$

where J_0 is the Bessel function of the first kind of order zero, and α_n are the roots of $J_0(\alpha) = 0$. The coefficients A_n are determined using the orthogonality of J_0 over $[0, R]$.

Bessel expansions are particularly relevant in solving axisymmetric heat equations, acoustic wave propagation in circular geometries, and in the thermal design of annular regenerator segments.

2.3.3 Convergence and truncation

While expansion series are mathematically exact in the limit of infinite terms, practical applications require truncation after a finite number of terms N. The truncation error depends on the smoothness of $f(x)$ and the choice of expansion basis. For Taylor series, the error is bounded by the next term in the series (via the Lagrange form of the remainder). For orthogonal expansions, convergence is typically faster for smooth functions due to exponential decay in higher-mode coefficients.

In engineering models of cryocoolers, balancing computational tractability with fidelity often requires retaining just a handful of modes, particularly in reduced-order models or when performing real-time system identification.

2.3.4 Applications in cryocooler modelling

Expansion series underpin many of the analytical techniques used throughout this book and serve as powerful tools for simplifying and solving complex mathematical models in cryogenic engineering. One of their primary uses is in solving linear and nonlinear ordinary and partial differential equations (ODEs and PDEs) that govern thermal and fluid dynamics in cryocooler subsystems. These equations often arise in modelling heat conduction, gas flow, and transient thermal responses, and expansion series allow for tractable, approximate solutions where closed-form expressions may not exist.

In regenerative cryocoolers, expansion series are particularly useful for decomposing temperature distributions into spatial harmonics. This enables the analysis of heat transport efficiency across the regenerator matrix, where periodic thermal waveforms can be described and interpreted in terms of their constituent modes. Moreover, in oscillating flow systems such as Stirling or pulse tube cryocoolers, expansion series facilitate the construction of frequency-domain models that represent the periodic pressure and displacement oscillations as sums of sinusoidal components. These representations help in understanding the phasing between pressure and flow, a critical factor in achieving high thermodynamic efficiency.

Expansion series also provide a means of developing approximate solutions for transient heat diffusion problems, where time-dependent temperature profiles evolve in response to boundary or initial condition changes. In such cases, orthogonal expansions, such as Fourier or Bessel series, allow for efficient numerical modelling and rapid convergence for smooth boundary profiles.

The selection of a suitable expansion basis is inherently problem-specific and depends on factors such as the geometry of the system, the nature of the boundary conditions, and the required accuracy of the solution. Throughout the remainder of this book, we will repeatedly make use of expansion series to derive analytical and semi-analytical solutions, examine system behaviour in both time and frequency domains, and build reduced-order models that enable fast and reliable performance predictions for a wide range of cryocooler architectures.

2.4 Differential equations

Differential equations are fundamental to the mathematical modelling of cryocoolers, as they govern the behaviour of physical quantities that vary with time, space, or both. In the context of cryogenic systems, these equations describe the evolution of temperature fields, gas pressures, fluid velocities, and structural displacements, among other variables, under the influence of thermodynamic and mechanical forces. This section introduces the core concepts of ordinary and partial differential equations, along with the methods most relevant for analysing cryocooler subsystems.

2.4.1 Ordinary differential equations

An ODE involves derivatives of a function with respect to a single independent variable, typically time t. These equations often arise in lumped-parameter models where spatial variation is negligible or has been averaged out. For example, the thermal response of a component with heat capacity C subject to a time-varying heat input $Q(t)$ is governed by the first-order ODE:

$$C\frac{\mathrm{d}T}{\mathrm{d}t} = Q(t), \tag{2.73}$$

where $T(t)$ is the temperature as a function of time. More complex systems, such as mass–spring–damper analogs of piston dynamics in Stirling machines, are modelled by second-order ODEs:

$$m\frac{\mathrm{d}^2x}{\mathrm{d}t^2} + c\frac{\mathrm{d}x}{\mathrm{d}t} + kx = F(t), \tag{2.74}$$

where $x(t)$ is displacement, m is mass, c is damping coefficient, k is spring constant, and $F(t)$ is an external forcing function such as oscillatory pressure.

The solution of ODEs involves initial conditions and may employ analytical methods (e.g. characteristic equations, Laplace transforms) or numerical techniques (e.g. Runge–Kutta methods), depending on linearity, order, and complexity.

2.4.2 Partial differential equations

PDEs arise when the function of interest depends on multiple independent variables, typically space and time. These are essential for modelling spatially distributed phenomena in cryocoolers, such as heat conduction along a regenerator or pressure wave propagation in gas columns. One canonical example is the one-dimensional heat equation:

$$\frac{\partial T}{\partial t} = \alpha \frac{\partial^2 T}{\partial x^2}, \tag{2.75}$$

where $T(x, t)$ is the temperature distribution, and α is the thermal diffusivity of the medium. PDEs of this type are solved subject to initial and boundary conditions, which may represent imposed temperatures, heat fluxes, or convective boundaries at the ends of a component.

Solutions to PDEs may be obtained through separation of variables, Green's functions, transform methods (e.g. Fourier or Laplace), or finite difference and finite element discretisation. These methods allow for high-fidelity thermal and fluid modelling, in particular in components with strong gradients or transient behaviour, such as cold ends, regenerators, and heat switches.

2.4.3 Linear and nonlinear behaviour

The governing equations of cryogenic systems are often linear under small perturbations or in idealised configurations. However, nonlinearities commonly arise in practical scenarios, such as temperature-dependent material properties, convective heat transfer coefficients, or geometric constraints in flexural suspensions. For example, the governing equation for pressure drop across a regenerator under oscillatory flow may involve nonlinear terms due to flow reversal and inertial effects.

Linear differential equations permit superposition and are amenable to frequency-domain analysis using phasors or transfer functions. Nonlinear equations typically require iterative or perturbative methods and may exhibit complex phenomena such as limit cycles or bifurcations, which are relevant in stability analysis and control.

2.4.4 Initial and boundary value problems

Differential equations in cryocooler modelling are typically posed either as initial value problems (IVPs), where the solution is propagated from known initial conditions, or as boundary value problems (BVPs), where the solution must satisfy constraints at multiple points. Transient thermal analyses, for instance, are often IVPs, while steady-state temperature profiles in a regenerator are BVPs requiring specified conditions at both the cold and warm ends.

The mathematical formulation and solution strategy depend on the classification of the differential equation (e.g. elliptic, parabolic, or hyperbolic in the case of PDEs) and the physical nature of the cryogenic subsystem being modelled.

2.4.5 Coupled systems and multiphysics models

In realistic cryocooler systems, differential equations rarely exist in isolation. Instead, they form coupled systems that describe interdependent physical processes, such as thermoacoustic interactions between gas dynamics and solid heat exchangers, or feedback loops between pressure oscillations and piston motion. These coupled models may involve mixed sets of ODEs and PDEs, and often require numerical simulation platforms for simultaneous solution.

Understanding the structure and interaction of these equations is critical for designing and optimising cryocooler performance. In later chapters, we will explore specific formulations of these systems and demonstrate how they can be simplified, linearised, or discretised to enable tractable analysis and control.

References

Franklin G F, Powell J D and Emami-Naeini A 1994 *Feedback Control of Dynamic Systems* 3rd edn (Englewood Cliffs, NJ: Prentice Hall)

Hayt W H, Kemmerly J E and Durbin S M 2019 *Engineering Circuit Analysis* 9th edn (New York: McGraw-Hill Education)

Oppenheim A V, Willsky A S and Young I T 1996 *Signals and Systems* 2nd edn (Englewood Cliffs, NJ: Prentice Hall)

Smith S 2017 *Electric Circuits and Phasor Analysis* (New York: Academic Press)

IOP Publishing

Hannah Rana

Chapter 3

Thermodynamics

3.1 First and second laws

Cryocooler performance is governed fundamentally by the classical laws of thermodynamics. In this section, we revisit the first and second laws, interpreting them from both a mathematical and physical standpoint, and framing them in the context of cryogenic systems.

Figure 3.1 presents a control-volume schematic of a generic cryocooler operating between a cold reservoir at temperature T_c and a heat rejector at T_h. The diagram summarises the fundamental thermodynamic interactions: input power $\dot{W}_{in}$, cooling power $\dot{Q}_c$, rejected heat $\dot{Q}_h$, and (optionally) recovered expansion work $\dot{W}_{exp}$. Internally, the system may store or deplete energy and entropy, represented by the time derivatives $\frac{d(mu)}{dt}$ and $\frac{d(ms)}{dt}$, respectively. The term $\dot{S}_{irr}$ accounts for entropy generated due to irreversibilities such as friction, pressure drops, and non-ideal heat exchange. This diagram serves as the foundation for the subsequent mathematical formulation of the first and second laws as applied to cryogenic systems.

3.1.1 The first law: energy conservation

The first law of thermodynamics, which formalises the conservation of energy, provides a fundamental constraint on all thermodynamic processes. For closed cryogenic systems, such as a complete cryocooler operating without mass exchange with its surroundings, the law may be written as

$$\frac{d}{dt}(mu) = \dot{Q}_c - \dot{Q}_0 - \dot{W}_{co} + \dot{W}_{exp}, \tag{3.1}$$

where m is the mass of the system, u is the specific internal energy, $\dot{Q}_c$ is the cooling load extracted from the cold reservoir at temperature T_c, $\dot{Q}_0$ is the heat rejected to the

doi:10.1088/978-0-7503-4826-3ch3

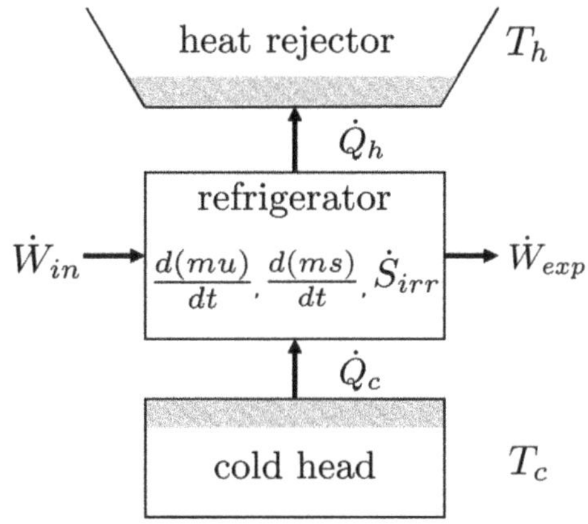

Figure 3.1. Control-volume representation of a cryocooler showing energy and entropy exchanges with its environment. The system absorbs cooling power $\dot{Q}_c$ from a cold reservoir at T_c, rejects heat $\dot{Q}_h$ to an ambient reservoir at T_h, and is driven by input work $\dot{W}_{in}$. Expansion work $\dot{W}_{exp}$ may be extracted depending on system architecture. Irreversible entropy production $\dot{S}_{irr}$ quantifies internal losses, while storage terms account for non-steady-state behaviour. (Image credit: Ray Radebaugh.)

ambient at T_h, and $\dot{W}_{co}$ and $\dot{W}_{exp}$ are the work input from the compressor and the expansion work output, respectively. Under steady-state operation, the internal energy of the system remains constant, and so $\frac{d}{dt}(mu) = 0$.

This energy balance becomes more intricate in cryocooler subsystems where mass crosses control surfaces. In such cases, for example through a regenerator or pulse tube, the system must be treated as an open thermodynamic element. Here, neglecting KE and PE, the first law for a control volume becomes

$$\dot{Q} + \sum_i \dot{m}_i h_i = \dot{W} + \sum_e \dot{m}_e h_e + \frac{d}{dt}(mu), \tag{3.2}$$

where $\dot{m}_i$ and $\dot{m}_e$ are the mass inflow and outflow rates, and h is the specific enthalpy. For steady-state operation, the storage term vanishes, simplifying the analysis of time-averaged behaviour.

This framework is crucial in regenerative cryocoolers, where oscillating helium gas flows across thermal interfaces and through components such as compressors, expanders, regenerators, and heat exchangers. In these systems, energy is cyclically stored and recovered within porous matrices or moving pistons. The concept of **availability**, defined as the maximum work obtainable as a fluid transitions from a given state to environmental conditions, emerges as a key quantity. The specific availability ψ for a given fluid stream, referenced to the surroundings at (T_0, p_0), is defined as

$$\psi = h - h_0 - T_0(s - s_0), \tag{3.3}$$

where h and s are the specific enthalpy and entropy, and the subscript 0 refers to properties at ambient conditions.

3.1.2 The second law: entropy and irreversibility

The second law of thermodynamics governs the directionality of energy transfer and introduces the concept of entropy as a measure of irreversibility. For a closed cryocooler, the entropy balance is

$$\frac{\mathrm{d}}{\mathrm{d}t}(ms) = \frac{\dot{Q}_{\mathrm{c}}}{T_{\mathrm{c}}} - \frac{\dot{Q}_0}{T_0} + \dot{S}_{\mathrm{irr}}, \tag{3.4}$$

where s is the specific entropy, and $\dot{S}_{\mathrm{irr}}$ is the rate of entropy generation due to irreversible processes such as heat conduction across finite temperature gradients, frictional pressure drops, and non-isentropic compression or expansion. Again, for steady-state systems, the storage term vanishes.

For open systems, the second law generalises to

$$\sum\nolimits_j \frac{\dot{Q}_j}{T_j} + \sum\nolimits_i \dot{m}_i s_i = \sum\nolimits_e \dot{m}_e s_e + \dot{S}_{\mathrm{irr}}, \tag{3.5}$$

where $\dot{Q}_j$ is the heat flow from a reservoir at temperature T_j, and the summation spans all inlets and outlets of the control volume.

The combination of the first and second laws for open systems yields an expression for the maximum useful work extractable from a flow stream, as in, the reversible work:

$$\dot{W}_{\mathrm{rev}} = \sum\nolimits_i \dot{m}_i(h_i - T_0 s_i) - \sum\nolimits_e \dot{m}_e(h_e - T_0 s_e), \tag{3.6}$$

which is diminished in real systems by the entropy production term

$$\dot{W} = \dot{W}_{\mathrm{rev}} - T_0\dot{S}_{\mathrm{irr}}. \tag{3.7}$$

This framework not only quantifies losses but also reveals design principles: to enhance cryocooler efficiency, one must reduce $\dot{S}_{\mathrm{irr}}$ through improved component design (for example minimisation of pressure drops, use of high-conductivity regenerator matrices, and isothermal compression/expansion). In steady-flow conditions, the reversible work per unit mass becomes

$$w_{\mathrm{rev}} = h_i - h_e - T_0(s_i - s_e), \tag{3.8}$$

which underpins performance calculations and optimisation strategies for real cryogenic systems.

3.1.3 Carnot limit and second-law efficiency

The theoretical upper bound for the performance of any cryogenic refrigeration system is set by the Carnot cycle. For a refrigerator operating between a cold temperature T_c and an ambient or warm temperature T_0, the ideal coefficient of performance (COP) is defined by

$$\mathrm{COP}_{\mathrm{Carnot}} = \frac{T_c}{T_0 - T_c}. \tag{3.9}$$

This expression represents a reversible process in which entropy generation is zero, and all heat and work interactions occur isothermally and without internal dissipation. Real cryocoolers, however, always deviate from this ideal due to irreversibilities distributed throughout the system, such as viscous dissipation, finite temperature gradients, imperfect regeneration, and non-isentropic compression and expansion.

To quantify how closely a real system approaches the Carnot limit, we define the second-law efficiency as

$$\eta_{\mathrm{II}} = \frac{\mathrm{COP}_{\mathrm{actual}}}{\mathrm{COP}_{\mathrm{Carnot}}}. \tag{3.10}$$

This efficiency provides a normalised metric that isolates internal entropy-producing effects from the fundamental temperature ratio of operation. It is especially useful when comparing systems operating at different temperature levels.

A more detailed understanding of inefficiencies in cryocoolers arises by integrating the first and second laws into a single expression for work input. For a control volume with heat flow $\dot{Q}_j$ from external reservoirs at temperatures T_j and with mass entering and exiting at states (h_i, s_i) and (h_e, s_e), respectively, the actual input power is

$$\dot{W} = \dot{W}_{\mathrm{rev}} + T_0 \dot{S}_{\mathrm{irr}}, \tag{3.11}$$

where $\dot{W}_{\mathrm{rev}}$ represents the reversible power, and the term $T_0 \dot{S}_{\mathrm{irr}}$ quantifies the **lost work**: the energetic penalty of irreversible entropy production within the system.

In practice, this lost work term dominates the deviation from ideal performance. It emerges from pressure losses, unrecouped acoustic power, enthalpy leaks, and thermal inefficiencies in exchangers and regenerators. The specific form of reversible work for a single stream is given by

$$w_{\mathrm{rev}} = (h_i - h_e) - T_0(s_i - s_e), \tag{3.12}$$

and for multiple flow paths and heat streams

$$\dot{W}_{\mathrm{rev}} = \sum\nolimits_i \dot{m}_i(h_i - T_0 s_i) - \sum\nolimits_e \dot{m}_e(h_e - T_0 s_e) + \sum\nolimits_j \left(1 - \frac{T_0}{T_j}\right)\dot{Q}_j. \tag{3.13}$$

The bracketed terms $(h - T_0 s)$ define the availability or **exergy** of a fluid stream, indicating its maximum useful work content relative to the surroundings. Availability analysis thus becomes a powerful tool for evaluating component-level inefficiencies, especially in regenerative cryocoolers, where work and entropy repeatedly cycle through storage and recovery phases.

An ideal system, free of entropy generation and with fully recovered expansion work, achieves the Carnot COP. In contrast, real systems experience reductions in performance that can be explicitly attributed to entropy generation and irreversible heat and pressure drops. The thermodynamic framework outlined here sets the stage for subsequent chapters, where we develop mathematical tools to model such losses and extract cycle-level efficiencies from component-level behaviours.

While the second law provides the directionality of thermodynamic processes, the concept of entropy generation, or irreversibility, quantifies the deviation of real systems from ideal behaviour. In cryocoolers, where efficiency is critical and operating margins are narrow, understanding and minimising entropy generation is essential to optimal design.

3.2 Enthalpy flows and energy accounting

A foundational step in cryocooler analysis involves understanding how energy moves through the system in the form of enthalpy flows, work transfer, and heat exchange. This forms the first law framework upon which more advanced entropy and exergy analyses are later constructed.

3.2.1 Definition and role of enthalpy

Enthalpy h serves as a practical thermodynamic property for flowing fluids because it encapsulates both the internal energy u and the flow work pv:

$$h = u + pv. \tag{3.14}$$

This relation is especially convenient in compressible fluid systems, where pressure–volume interactions are significant, and separates neatly in enthalpy-based energy accounting.

In addition, the total energy content per unit mass, including kinetic and potential terms, is given as

$$e_{\text{total}} = u + \frac{v^2}{2} + gz. \tag{3.15}$$

However, in many cryocooler components, changes in kinetic energy and elevation are negligible, and enthalpy captures the dominant modes of energy transfer.

3.2.2 Steady-flow energy balance

The steady-flow energy balance for an open system (control volume) is given by

$$\dot{Q} - \dot{W} = \dot{m}(h_{\text{out}} - h_{\text{in}}), \tag{3.16}$$

where $\dot{Q}$ is the rate of heat transfer into the control volume, $\dot{W}$ is the rate of work done *by* the control volume (positive if output), $\dot{m}$ is the mass flow rate through the control volume, and h_{in} and h_{out} are the specific enthalpies at the inlet and outlet, respectively.

Cryocooler subsystems often feature multiple inlet and outlet streams, such as regenerators with bi-directional flow or buffer volumes with thermal interaction. For these, the energy balance generalises to

$$\sum_{\text{inlets}} \dot{m}_i h_i - \sum_{\text{outlets}} \dot{m}_j h_j = \dot{Q} - \dot{W}, \tag{3.17}$$

where the sums account for all entering and exiting streams. This form enables local accounting of energy transfer within complex cryocooler architectures.

3.2.3 Time-averaged enthalpy flow in oscillating systems

Cryocoolers operate with strongly oscillatory mass flow, particularly in regenerative types. Therefore, enthalpy flow is most meaningfully treated as a time-averaged quantity over one or more cycles:

$$\langle \dot{H} \rangle = \langle \dot{m}(t) h(t) \rangle, \tag{3.18}$$

which represents the net energy transfer associated with fluid flow into or out of a control volume. This averaged enthalpy flow is critical for computing the effective cooling power at the cold heat exchanger. The net cooling rate is then

$$\dot{Q}_c = \langle \dot{m} \rangle (h_{\text{in}} - h_{\text{out}}), \tag{3.19}$$

where h_{in} and h_{out} refer to the specific enthalpies of the working fluid entering and exiting the cold stage.

3.2.4 Application to compressor and regenerator analysis

The work input to the compressor can be estimated from the enthalpy rise of the working gas. Under adiabatic conditions, where heat transfer is negligible, the required shaft power is approximated by

$$\dot{W}_{\text{in}} = \dot{m}(h_{\text{out}} - h_{\text{in}}), \tag{3.20}$$

capturing the net increase in energy per unit mass of the fluid. This formulation is particularly useful in comparing theoretical work input to measured electrical input power.

The effectiveness of the regenerator, a key component in maintaining thermodynamic efficiency, can also be analysed via enthalpy. A common definition compares the enthalpy change of the returning gas with the ideal enthalpy swing between hot and cold ends:

$$\varepsilon_{\text{regen}} = \frac{\langle h_{\text{return}} \rangle - h_{\text{min}}}{h_{\text{max}} - h_{\text{min}}}, \tag{3.21}$$

where h_{max} and h_{min} are the enthalpies at the hot and cold ends of the regenerator, respectively, and $\langle h_{\text{return}} \rangle$ is the time-averaged enthalpy of the gas leaving the cold end and re-entering the regenerator toward the hot end. A high regenerator effectiveness indicates efficient heat recovery and minimal entropy generation.

3.2.5 Wall heat transfer and generalised enthalpy expressions

In cases where heat is added from solid boundaries such as a heat exchanger or housing wall, this thermal input appears as a distributed heat flux term. The net heat transfer rate is given by

$$\dot{Q}_{\text{wall}} = \int_A q'' \, \mathrm{d}A, \tag{3.22}$$

where q'' is the local heat flux and A is the surface area of the thermal interface. This contribution must be included in the first law energy balance for realistic modelling of component boundaries.

In systems where the working gas is ideal and has a temperature-dependent specific heat capacity, enthalpy can be computed from temperature as

$$h(T) = \int_{T_{\text{ref}}}^{T} c_{\text{p}}(T') \, \mathrm{d}T', \tag{3.23}$$

where T_{ref} is a reference temperature and $c_{\text{p}}(T')$ is the specific heat at constant pressure.

3.2.6 Applications and link to exergy analysis

Beyond energy conservation, enthalpy flows provide critical insights into cryocooler performance evaluation and set the stage for exergy-based assessments. The same enthalpy differences used to compute cooling capacity and work input also factor into the specific exergy.

Thus, enthalpy analysis not only enables accurate energy tracking throughout a cryocooler system but also provides the essential inputs to more advanced evaluations of thermodynamic irreversibility, entropy generation, and system-level efficiency.

3.3 Entropy generation and irreversibility

While the second law provides the directionality of thermodynamic processes, the concept of entropy generation, or, 'irreversibility', quantifies the deviation of real systems from ideal behaviour. In cryocoolers, where efficiency is critical and operating margins are narrow, understanding and minimising entropy generation is essential to optimal design.

3.3.1 Sources of entropy generation

Entropy generation, denoted $\dot{S}_{\text{irr}}$, arises from irreversibilities within a thermodynamic system, mechanisms by which energy is dissipated and rendered unavailable for useful work. In cryocoolers, where thermal energy must be moved efficiently from a low-temperature source to a higher-temperature sink, these irreversibilities represent a direct penalty to performance. Although some degree of entropy production is inevitable in practical systems, its minimisation is central to achieving high second-law efficiency and approaching the Carnot limit.

One of the most prevalent sources of entropy generation in cryogenic systems is the presence of thermal gradients. Whenever heat flows across a finite temperature difference, as occurs in heat exchangers and regenerators, entropy is generated. Ideal heat transfer would occur isothermally, allowing energy exchange without entropy production. However, in real components, temperature differences are necessary to drive heat across finite conductances. The local entropy generation due to heat conduction across a temperature gradient is given by

$$\dot{S}_{\text{cond}} = \int_A \frac{\vec{q} \cdot \vec{n}}{T} \text{d}A, \tag{3.24}$$

where $\vec{q}$ is the heat flux vector, $\vec{n}$ is the unit normal vector to the surface, and T is the local temperature. In the case of one-dimensional conduction, this simplifies to

$$\dot{S}_{\text{cond}} = \int_{x_1}^{x_2} \frac{kA}{T^2} \left(\frac{\text{d}T}{\text{d}x} \right)^2 \text{d}x, \tag{3.25}$$

where k is the thermal conductivity, A is cross-sectional area, and $\text{d}T/\text{d}x$ is the temperature gradient.

For example, in a regenerator, the gas oscillates back and forth through a matrix that stores and releases heat, and the temperature profile of the gas is never in perfect synchrony with that of the matrix. These mismatches result in net entropy production over each cycle.

Another major contributor is viscous dissipation, often accompanied by pressure drops across the working fluid pathway. As helium or another cryogenic gas is forced through narrow channels, screens, or packed beds, particularly within the regenerator, the viscous interaction between the gas and solid boundaries leads to irreversible losses. The entropy generation due to viscous flow in a duct is given by

$$\dot{S}_{\text{visc}} = \int_V \frac{\Phi}{T} \text{d}V, \tag{3.26}$$

where Φ is the viscous dissipation function, which for laminar flow in a pipe can be approximated as

$$\Phi = \mu \left(\frac{du}{dy} \right)^2, \tag{3.27}$$

with μ being the dynamic viscosity and $\text{d}u/\text{d}y$ the velocity gradient.

These frictional effects convert mechanical energy into internal energy (heat) and are a function of flow velocity, geometry, and surface roughness. In oscillating systems, such as pulse tube or Stirling-type cryocoolers, these effects are cyclic and cumulative, and even small pressure drops can compound to significant entropy generation over time.

Non-isentropic compression and expansion processes are another important source. Ideally, compression and expansion of the working gas should occur isentropically, without heat exchange and without entropy change. The entropy change for a real polytropic process from state 1 to 2 can be expressed as

$$\Delta s = c_p \ln\left(\frac{T_2}{T_1}\right) - R \ln\left(\frac{p_2}{p_1}\right), \tag{3.28}$$

where c_p is the specific heat at constant pressure, R is the gas constant, and (p_1, T_1) and (p_2, T_2) are the initial and final pressures and temperatures, respectively. Any deviation from $\Delta s = 0$ implies irreversibility.

In real compressors and expanders, however, factors such as finite piston speed, mechanical friction, gas leakage, and imperfect insulation lead to departures from the isentropic ideal. These processes may also involve heat exchange with the surroundings or the cylinder walls, further increasing entropy. The result is additional input power required to achieve a given temperature gradient, reducing the overall coefficient of performance, which is defined as

$$\text{COP} = \frac{\dot{Q}_c}{\dot{W}}. \tag{3.29}$$

A further mechanism of irreversibility arises from mixing and turbulence within the gas flow, particularly in geometrically complex regions such as phase shifters, flow impedances, and junctions. In some cryocooler designs, sharp transitions, sudden expansions, or vortex-prone geometries may induce localised turbulence or eddies. These flows tend to dissipate kinetic energy and create microscale temperature and velocity gradients that enhance entropy production. For compressible turbulent flows, the entropy generation per unit mass may be approximated as

$$\dot{s}_{\text{turb}} \approx \frac{\varepsilon}{T}, \tag{3.30}$$

where ε is the turbulent kinetic energy dissipation rate.

Unlike the reversible laminar flow assumed in ideal models, these turbulent effects are inherently dissipative and difficult to recover.

Each of these entropy-generating mechanisms contributes to the overall irreversibility of the cycle and must be offset by an increased input of mechanical or electrical work. From a system-level perspective, $\dot{S}_{\text{irr}}$ quantifies how far a given cryocooler departs from the ideal of reversible operation. In the limiting case where all processes are perfectly reversible, we have

$$\dot{S}_{\text{irr}} = 0, \tag{3.31}$$

and the cryocooler achieves its theoretical maximum performance. In practice, however, the focus shifts to minimising each contributor to entropy generation through careful design of components, optimisation of flow dynamics, and thermal management strategies.

3.3.2 Quantifying irreversibility: combined laws

To rigorously account for the performance degradation caused by entropy generation, we turn to the combined application of the first and second laws of thermodynamics for open systems. When applied under steady-state conditions, where no accumulation of energy or entropy occurs over time, these laws can be used together to isolate the thermodynamic cost of irreversibility in cryocooler operation.

The key result is a decomposition of the total power input $\dot{W}$ into two terms:

$$\dot{W} = \dot{W}_{\text{rev}} + T_0\dot{S}_{\text{irr}}, \tag{3.32}$$

where $\dot{W}_{\text{rev}}$ represents the minimum or *reversible* work required to achieve the desired cooling performance, and $T_0\dot{S}_{\text{irr}}$ quantifies the additional input power (or, *lost work*) that must be supplied to overcome irreversibilities within the system. The ambient temperature T_0 serves as the reference environment, typically the temperature of the heat sink or surrounding atmosphere.

This relationship provides a powerful framework for evaluating cryocooler efficiency. The reversible work $\dot{W}_{\text{rev}}$ is dictated by the thermodynamic boundary conditions, such as the temperatures T_c and T_0, and the desired cooling power $\dot{Q}_c$, and represents a theoretical minimum that cannot be improved upon regardless of engineering effort. In the case of a simple refrigeration cycle, the reversible work can be expressed as

$$\dot{W}_{\text{rev}} = \dot{Q}_c\left(\frac{T_0 - T_c}{T_c}\right), \tag{3.33}$$

which follows directly from the ideal Carnot relation.

In contrast, the entropy generation term $\dot{S}_{\text{irr}}$ encapsulates the consequences of all internal losses and inefficiencies, including pressure drops, thermal resistances, non-ideal fluid flow, and non-isentropic compression or expansion. Reducing this term is thus the key to improving real-world performance. A more general expression for entropy generation in control-volume systems with heat and mass exchange is given by

$$\dot{S}_{\text{irr}} = \sum\left(\frac{\dot{Q}_i}{T_i}\right)_{\text{in}} - \sum\left(\frac{\dot{Q}_j}{T_j}\right)_{\text{out}} + \dot{m}_{\text{in}}s_{\text{in}} - \dot{m}_{\text{out}}s_{\text{out}}, \tag{3.34}$$

where the summations are over all heat inputs and outputs at their respective temperatures.

In regenerative cryocoolers, this decomposition is especially relevant due to the inherently oscillatory nature of mass flow, pressure, and enthalpy. Unlike steady-flow systems, regenerative devices involve cyclic processes in which energy is temporarily stored in components such as the regenerator matrix or the gas's thermal capacity and later recovered. The efficiency of this storage-and-recovery process depends critically on phase relationships between pressure, flow, and temperature. The enthalpy flux of the oscillating gas can be expressed as a time-dependent function:

$$\dot{H}(t) = \dot{m}(t)h(t) = \rho(t)u(t)Ah(t), \tag{3.35}$$

where $u(t)$ is the oscillating velocity, A is the cross-sectional area, and $h(t)$ is the instantaneous specific enthalpy.

When these quantities are optimally phased, the regenerator can transfer heat with minimal entropy production. However, any mismatch in timing or amplitude, such as flow reversal occurring before sufficient heat transfer is complete, leads to increased irreversibility and reduced effectiveness.

Moreover, this thermodynamic formalism allows designers to define performance metrics that isolate specific sources of degradation. For example, the ratio

$$\eta_{\mathrm{II}} = \frac{\dot{W}_{\mathrm{rev}}}{\dot{W}} \tag{3.36}$$

defines the second-law efficiency, providing a normalised measure of how much closer a real system operates to the ideal. When $\dot{S}_{\mathrm{irr}}$ is small, this ratio approaches unity, and the system operates with high thermodynamic quality. Conversely, large entropy generation reflects poor utilisation of the input work and often indicates design features that need revision, whether it be excessive pressure drop in a regenerator or thermal mismatch at a heat exchanger.

In practical terms, the expression $\dot{W} = \dot{W}_{\mathrm{rev}} + T_0\dot{S}_{\mathrm{irr}}$ becomes an essential diagnostic and design tool. By quantifying both the theoretical and actual work requirements, engineers can assess not only how efficient a cryocooler is, but also why it performs the way it does. The goal in design optimisation then becomes twofold: to reduce $\dot{S}_{\mathrm{irr}}$ through improved component performance, and to ensure that $\dot{W}_{\mathrm{rev}}$ is met with as little surplus input power as possible.

3.3.3 Availability and the role of entropy flow

Exergy quantifies the maximum amount of useful work that can be extracted from a system as it transitions from its current state to equilibrium with a defined environment, typically taken to be a large reservoir at temperature T_0 and pressure p_0. Unlike energy, which is conserved according to the first law, exergy is not conserved: it is *degraded* through irreversible processes such as heat transfer across finite temperature differences, viscous dissipation, mixing, and friction. As such, exergy provides a direct measure of both the capacity to perform work and the reduction of that capacity due to entropy generation.

The total exergy flow rate associated with a mass flow $\dot{m}$ is

$$\dot{X} = \dot{m}\psi. \tag{3.37}$$

In the context of cryocoolers, where working fluids such as helium oscillate between thermal reservoirs and components, the concept of exergy is particularly instructive. The enthalpy term in the exergy expression accounts for the total energy content of the fluid (thermal and mechanical), while the entropy term $T_0 s$ reflects the portion of this energy that is unavailable for work. Subtracting the environmental baseline ensures that the exergy reflects only the excess work potential relative to ambient conditions.

The flow of entropy itself, expressed as $\dot{m}s$, plays a central role in the degradation of exergy. Even when a fluid stream carries substantial energy, a high-entropy content implies that much of that energy is disordered and inaccessible for conversion into useful work. The term $T_0\dot{m}s$ quantifies this unavailable portion.

As a result, systems that permit significant entropy transport, particularly from cold to warm regions, suffer from elevated exergy losses. For instance, if high-entropy gas exits the regenerator and enters the pulse tube, it reduces the net exergy available for producing cooling at the cold tip.

The second law provides a direct link between entropy generation and exergy degradation. The exergy destruction rate due to internal irreversibility is given by

$$\dot{X}_{\text{dest}} = T_0 \dot{S}_{\text{irr}}. \tag{3.38}$$

This quantity represents the lost work potential per unit time caused by non-ideal processes within the cryocooler. Minimising $\dot{S}_{\text{irr}}$ thus directly corresponds to minimising exergy destruction, and hence maximising the useful output from a given input of mechanical or electrical power.

Furthermore, a control-volume version of the exergy balance can be written as

$$\sum \dot{X}_{\text{in}} - \sum \dot{X}_{\text{out}} = \dot{W}_{\text{useful}} + T_0 \dot{S}_{\text{irr}}, \tag{3.39}$$

where $\dot{W}_{\text{useful}}$ is the net useful work delivered by the control volume, and the terms on the left-hand side represent the net exergy carried by the incoming and outgoing fluid streams and heat fluxes. This expression allows designers to quantify and localise where in the system exergy is being consumed, whether it be across heat exchangers, within regenerators, or through non-isentropic compression and expansion.

Because the $T_0 s$ term scales with both entropy and ambient temperature, exergy degradation becomes more pronounced at higher environmental temperatures. Systems operating in warm surroundings must therefore be even more careful in minimising entropy generation. Otherwise, a given amount of internal entropy production leads to a larger exergy loss:

$$(\text{greater } T_0 \Rightarrow \text{more severe } \dot{X}_{\text{dest}} \text{ for a given } \dot{S}_{\text{irr}}). \tag{3.40}$$

In summary, exergy analysis enables engineers to interpret entropy generation not only as a thermodynamic inefficiency but as a direct loss of work potential. It captures the quality of energy in addition to its quantity and provides a rigorous framework for identifying and minimising the performance penalties associated with irreversibility. By embedding exergy accounting into the analysis of each component, cryocooler designers can better understand how material choices, geometry, flow regimes, and phasing influence overall system performance. Ultimately, minimising exergy destruction becomes the thermodynamic imperative that drives innovation toward high-efficiency, near-Carnot cryogenic systems.

3.3.4 Entropy balance and practical implications

The entropy balance equation provides a rigorous quantitative foundation for evaluating irreversibility in cryocoolers. Under steady-state conditions, where there is no net accumulation of entropy within the system over time, the second law of thermodynamics reduces to a balance between the incoming and outgoing entropy fluxes. For a control volume encompassing the cryocooler, the time-invariant form of the entropy balance is given by

$$\sum \frac{\dot{Q}_{\text{in}}}{T_{\text{in}}} - \sum \frac{\dot{Q}_{\text{out}}}{T_{\text{out}}} + \sum_i \dot{m}_i s_i - \sum_e \dot{m}_e s_e = \dot{S}_{\text{irr}}, \tag{3.41}$$

where the first two terms account for entropy carried by heat transfer at boundary temperatures T_{in} and T_{out}, and the last two terms represent entropy carried by mass entering and leaving the control volume. In cryocoolers operating in closed-loop or oscillatory cycles, the net mass exchange is typically zero over a cycle, allowing this expression to simplify.

In the common case where the system operates between two thermal reservoirs, a cold one at temperature T_{c} and a hot one at T_{h}, and heat flows $\dot{Q}_{\text{c}}$ and $\dot{Q}_{\text{h}}$ are exchanged with them, the entropy balance simplifies to

$$\frac{\dot{Q}_{\text{h}}}{T_{\text{h}}} - \frac{\dot{Q}_{\text{c}}}{T_{\text{c}}} = \dot{S}_{\text{irr}}. \tag{3.42}$$

This relation defines the fundamental lower bound on entropy production within the system. For an ideal, perfectly reversible cryocooler, $\dot{S}_{\text{irr}} = 0$, and we recover the Carnot condition:

$$\frac{\dot{Q}_{\text{h}}}{T_{\text{h}}} = \frac{\dot{Q}_{\text{c}}}{T_{\text{c}}}. \tag{3.43}$$

In real systems, any deviation from this balance reflects thermodynamic losses and irreversible processes occurring internally.

This formulation emphasises the sensitivity of cryocooler performance to even small internal losses. Since entropy generation increases with nonzero temperature gradients, pressure drops, and phase mismatches, seemingly minor sources of irreversibility, such as a pressure loss ΔP across the regenerator or heat exchanger, can lead to measurable increases in $\dot{S}_{\text{irr}}$. For example, the entropy produced due to viscous dissipation in a flow passage is approximately:

$$\dot{S}_{\text{irr,flow}} = \frac{\dot{m}\Delta P}{\rho T}, \tag{3.44}$$

where $\dot{m}$ is the mass flow rate, ρ is the fluid density, and T is the local temperature. As T_{c} approaches cryogenic values, the penalty associated with a given $\dot{S}_{\text{irr}}$ becomes more severe due to the increased entropy content per unit of cooling power:

$$\frac{\dot{Q}_{\text{c}}}{T_{\text{c}}} \to \infty \quad \text{as} \quad T_{\text{c}} \to 0. \tag{3.45}$$

This nonlinear scaling underscores the thermodynamic difficulty of achieving deep cryogenic cooling without significant increases in input work.

The entropy balance also serves as a diagnostic and optimisation tool during design and testing. Applied across the entire system, it offers a global snapshot of entropy production and efficiency. However, it can also be localised to individual components using a control-volume approach. In unsteady systems, such as Stirling or pulse tube cryocoolers, the instantaneous application of conservation laws is complicated by oscillating flows of energy and entropy. Nonetheless, by averaging over one or more complete cycles, we recover meaningful steady-state representations of heat flow $\bar{\dot{Q}}$, entropy flow $\bar{\dot{S}}$, and acoustic power $\overline{\dot{W}_{ac}}$.

For instance, consider the regenerator, where heat is transferred periodically between the gas and the matrix. A time-averaged entropy flow $\overline{\dot{m}s(x)}$ that increases monotonically along the flow direction (from cold to hot end) is indicative of irreversible losses. This gradient may be expressed as

$$\frac{\mathrm{d}}{\mathrm{d}x}(\overline{\dot{m}s}) = \dot{s}_{\mathrm{gen}}(x) > 0, \tag{3.46}$$

where $\dot{s}_{\mathrm{gen}}(x)$ is the local volumetric entropy generation rate. A similar formulation holds in the pulse tube, where entropy generation may arise from acoustic streaming, phase lag between flow and pressure, or thermal boundary layers. When integrated over the entire component, the total entropy generation becomes a key figure of merit for that segment of the system.

Additionally, the entropy balance provides a pathway to estimate the thermodynamic cost of irreversibility. The lost work associated with internal entropy generation is given by

$$\dot{W}_{\mathrm{lost}} = T_0 \dot{S}_{\mathrm{irr}}, \tag{3.47}$$

where T_0 is the ambient temperature of the surroundings. This lost work must be supplied by the compressor or input driver and directly contributes to increased energy consumption. The second-law efficiency can thus be evaluated as

$$\eta_{\mathrm{II}} = \frac{\dot{W}_{\mathrm{rev}}}{\dot{W}_{\mathrm{actual}}} = \frac{\dot{Q}_{\mathrm{c}}(T_{\mathrm{h}}/T_{\mathrm{c}} - 1)}{\dot{Q}_{\mathrm{h}} - \dot{Q}_{\mathrm{c}}}. \tag{3.48}$$

This formulation cleanly separates ideal thermodynamic requirements from actual energetic expenditures and provides a target for system improvement.

Ultimately, the entropy balance equation encapsulates the second law's governing influence on cryocooler design. While the first law sets the stage by conserving total energy, the second law determines how that energy can be usefully transformed. By embedding entropy accounting into the analysis of each subsystem; regenerator, pulse tube, compressor, heat exchangers; engineers can map where and how irreversibilities arise, quantify their impact, and design mitigation strategies. When paired with exergy and phasor-based methods, entropy balance becomes a central analytical tool for driving cryocooler systems toward their fundamental thermodynamic limits.

Worked example: entropy generation in a cryocooler

Consider a regenerative cryocooler operating between a cold reservoir at $T_{\mathrm{c}} = 80$ K and an ambient environment at $T_{\mathrm{h}} = 300$ K. The device extracts a steady cooling load of $\dot{Q}_{\mathrm{c}} = 5$ W from the cold end and rejects $\dot{Q}_{\mathrm{h}} = 25$ W to the surroundings. We wish to calculate the internal entropy generation rate $\dot{S}_{\mathrm{irr}}$ and assess how far the system is operating from the reversible limit.

From the steady-state entropy balance, we use

$$\dot{S}_{\mathrm{irr}} = \frac{\dot{Q}_{\mathrm{h}}}{T_{\mathrm{h}}} - \frac{\dot{Q}_{\mathrm{c}}}{T_{\mathrm{c}}}. \tag{3.49}$$

Substituting the known values:

$$\frac{\dot{Q}_{\mathrm{h}}}{T_{\mathrm{h}}} = \frac{25\ \mathrm{W}}{300\ \mathrm{K}} = 0.0833\ \mathrm{WK}^{-1},$$

$$\frac{\dot{Q}_{\mathrm{c}}}{T_{\mathrm{c}}} = \frac{5\ \mathrm{W}}{80\ \mathrm{K}} = 0.0625\ \mathrm{WK}^{-1}.$$

Thus, the internal entropy generation rate is

$$\dot{S}_{\mathrm{irr}} = 0.0833 - 0.0625 = 0.0208\ \mathrm{WK}^{-1}. \tag{3.50}$$

This result means that the cryocooler generates approximately 20.8 mW K^{-1} of entropy internally due to irreversibilities. To assess the thermodynamic penalty of this entropy generation, we compute the associated lost work:

$$\dot{W}_{\mathrm{lost}} = T_{\mathrm{h}}\dot{S}_{\mathrm{irr}} = 300\ \mathrm{K} \times 0.0208\ \mathrm{WK}^{-1} = 6.25\ \mathrm{W}. \tag{3.51}$$

This indicates that, in addition to the ideal work required to remove the 5 W of cooling power at 80 K, an extra 6.24 W of input power must be supplied solely to compensate for internal inefficiencies.

The second-law (exergy) efficiency of the system can now be calculated by comparing the minimum reversible work input to the actual work input. The reversible work is given by

$$\dot{W}_{\mathrm{rev}} = \dot{Q}_{\mathrm{c}}\left(\frac{T_{\mathrm{h}}}{T_{\mathrm{c}}} - 1\right) = 5\ \mathrm{W}\left(\frac{300}{80} - 1\right) = 13.75\ \mathrm{W}. \tag{3.52}$$

The actual work input is the difference between heat rejected and absorbed:

$$\dot{W}_{\mathrm{actual}} = \dot{Q}_{\mathrm{h}} - \dot{Q}_{\mathrm{c}} = 25\ \mathrm{W} - 5\ \mathrm{W} = 20\ \mathrm{W}. \tag{3.53}$$

The second-law efficiency is therefore

$$\eta_{\mathrm{II}} = \frac{\dot{W}_{\mathrm{rev}}}{\dot{W}_{\mathrm{actual}}} = \frac{13.75}{20} = 0.6875. \tag{3.54}$$

This shows that the cryocooler is operating at approximately 68.75% of its ideal thermodynamic potential. While this performance is fairly good, the example highlights that even modest entropy generation can result in significant lost work, especially at low cold-end temperatures. Efficient cryocooler design must therefore aim to reduce $\dot{S}_{\mathrm{irr}}$ through careful control of flow dynamics, thermal coupling, and component-level losses.

3.3.5 Design strategies to reduce $\dot{S}_{\text{irr}}$

The thermodynamic analyses in previous sections reveal that irreversibilities, quantified by the entropy generation rate $\dot{S}_{\text{irr}}$, directly reduce cryocooler efficiency by increasing the required input power beyond the theoretical minimum. While it is thermodynamically impossible to eliminate entropy generation entirely in any real system, careful engineering can reduce its magnitude, localise it to less critical areas, and optimise system performance. This section outlines key strategies for minimising $\dot{S}_{\text{irr}}$ and supporting high second-law efficiency.

A central design consideration is the optimisation of the regenerator, which is responsible for transferring heat between the working fluid and a thermal matrix during the oscillating flow. Entropy generation in a regenerator arises primarily from two sources: imperfect heat exchange across finite temperature gradients and viscous dissipation due to pressure drop. For an ideal regenerator, the time-averaged entropy flow $\bar{\dot{S}}$ would remain constant along its length. In a real regenerator, the local rate of entropy generation can be modelled as

$$\dot{S}_{\text{irr,reg}} = \int_0^L \left[\frac{kA}{T^2}\left(\frac{\mathrm{d}T}{\mathrm{d}x}\right)^2 + \frac{\mu}{\rho T}\left(\frac{\mathrm{d}u}{\mathrm{d}x}\right)^2 \right] \mathrm{d}x, \tag{3.55}$$

where k is the thermal conductivity of the gas, A is the flow cross-sectional area, μ is the dynamic viscosity, ρ is the gas density, and $u(x)$ is the axial velocity profile. The first term corresponds to entropy generated by heat conduction down a temperature gradient, and the second represents entropy produced by viscous shear.

Reducing the thermal contribution requires minimising temperature gradients $\frac{\mathrm{d}T}{\mathrm{d}x}$, which can be achieved by selecting regenerator materials with high volumetric heat capacity and optimising matrix geometry for efficient thermal contact. The viscous term is minimised by lowering flow velocities or increasing hydraulic diameter, thus reducing velocity gradients $\frac{\mathrm{d}u}{\mathrm{d}x}$ and pressure drops. Since the pressure drop across the regenerator, ΔP, contributes to entropy generation via

$$\dot{S}_{\text{irr,flow}} = \frac{\dot{m}\Delta P}{\rho T}, \tag{3.56}$$

a small ΔP is desirable to keep losses minimal. Here, $\dot{m}$ is the mass flow rate through the regenerator, and T is the mean gas temperature. This shows that entropy generation scales linearly with flow rate and pressure drop but inversely with temperature, making low-temperature components more sensitive to inefficiencies.

Beyond the regenerator, the broader cryocooler layout must be designed to minimise pressure drops in all internal flow paths, including pulse tubes, heat exchangers, and connecting passages. Total entropy generation due to pressure losses can be approximated by

$$\dot{S}_{\text{irr,tot}} \approx \sum_i \frac{\dot{m}_i \Delta P_i}{\rho_i T_i}, \tag{3.57}$$

summed over all components i. This cumulative view emphasises the importance of uniform optimisation: a poorly designed restriction anywhere in the system can dominate the total $\dot{S}_{irr}$. Pressure drop scaling laws show that:

$$\Delta P \propto \frac{L\dot{m}^2}{D^5}, \tag{3.58}$$

where L is the channel length and D is its hydraulic diameter. This relationship motivates the use of large-diameter passages with streamlined geometries where possible.

In oscillatory cryocoolers, entropy generation also arises from improper phasing between pressure and flow oscillations. Ideally, energy transfer should occur coherently, with the pressure and mass flow phasors closely aligned. Misalignment leads to energy being expended in moving the fluid against unfavourable pressure gradients, which is then dissipated. The time-averaged acoustic (PV) power flow, a proxy for recoverable work, is given by

$$\overline{\dot{W}_{\text{acoustic}}} = \frac{1}{2}\text{Re}\{\tilde{P}\tilde{U}^*\} = \frac{1}{2}|\tilde{P}||\tilde{U}|\cos\theta, \tag{3.59}$$

where $\tilde{P}$ and $\tilde{U}$ are the complex phasors of pressure and volumetric flow rate, and θ is their phase difference. A smaller phase angle θ maximises useful work transfer and minimises dissipation. This motivates careful tuning of inertance tubes, compliance volumes, and valve timing to establish favourable phasing.

Thermal gradients are another critical source of entropy generation, especially across heat exchangers. The local entropy generation rate due to heat transfer across a finite temperature difference ΔT is approximately

$$\dot{S}_{\text{irr,thermal}} = \frac{\dot{Q}\Delta T}{T_{\text{avg}}^2}, \tag{3.60}$$

where T_{avg} is the average temperature of the interface. As ΔT increases or T_{avg} decreases, the entropy generation grows rapidly. This highlights the importance of designing heat exchangers that promote isothermal heat transfer. High-conductivity materials, distributed flow architectures, and large surface areas help reduce temperature differentials and suppress associated entropy generation.

Overall, while entropy generation cannot be eradicated, its impact can be significantly reduced through deliberate thermodynamic design. Regenerators must balance high heat transfer rates with low pressure drops. Flow paths must be shaped to reduce frictional losses. Phase alignment must be tuned to preserve acoustic power, and heat exchangers must be configured to maintain near-isothermal interfaces. These strategies collectively minimise the lost work

$$\dot{W}_{\text{lost}} = T_0\dot{S}_{\text{irr}}, \tag{3.61}$$

and elevate second-law efficiency

$$\eta_{\text{II}} = \frac{\dot{W}_{\text{rev}}}{\dot{W}_{\text{actual}}} = \frac{\dot{Q}_c(T_h/T_c - 1)}{\dot{Q}_h - \dot{Q}_c}. \tag{3.62}$$

The effectiveness of any cryocooler hinges on how well it controls and confines entropy generation. The closer the system operates to its ideal reversible configuration, the smaller the entropy production per watt of cooling, and the more competitive the device becomes in power-limited or space-constrained applications.

3.4 Thermodynamic cycles

Cryocoolers operate on thermodynamic cycles that transform input work into cooling power by transporting heat from a low-temperature source to a higher-temperature sink. These cycles are inherently governed by the first and second laws of thermodynamics, and their performance depends critically on both the idealised process steps and the real-world irreversibilities that degrade them. In this section, we explore the foundational thermodynamic cycles that underpin cryocooler operation, their idealised efficiency limits, and the modifications required for practical implementation.

3.4.1 Ideal reversible cycles

The Carnot cycle provides the theoretical upper bound on cryocooler performance. It consists of four reversible steps. First, heat is absorbed isothermally at the cold reservoir temperature T_c, during which the working fluid extracts thermal energy from the load without changing its temperature. This is followed by an isentropic (i.e. reversible adiabatic) compression process that raises the temperature of the fluid from T_c to the ambient temperature T_0, increasing its pressure and internal energy without heat exchange.

Next, the cycle includes an isothermal heat rejection at temperature T_0, where the working fluid releases its energy to the environment while maintaining a constant temperature. Finally, the working fluid undergoes isentropic expansion from T_0 back to T_c, during which it performs work and reduces its internal energy, thereby completing the cycle.

For such an idealised cycle, the COP for refrigeration is given by

$$\mathrm{COP}_{\mathrm{Carnot}} = \frac{T_c}{T_0 - T_c}, \tag{3.63}$$

where T_c is the cold reservoir temperature and T_0 is the ambient (sink) temperature. This expression defines the maximum theoretical efficiency of any cryocooler operating between these two thermal boundaries.

However, the Carnot cycle is not directly realisable in practice due to its requirement for perfectly reversible processes and zero entropy generation. Real devices must approximate this limit using implementable processes that inherently involve some degree of irreversibility.

3.4.2 Real cryocooler cycles

Practical cryocoolers implement thermodynamic cycles that retain favourable aspects of idealised models while accommodating realistic constraints such as finite heat transfer, mechanical losses, pressure drops, and non-ideal working fluids. Among the most widely used cycles in cryogenic refrigeration are the Stirling, Gifford–McMahon (GM), pulse tube, and Joule–Thomson (JT) cycles.

3.4.2.1 Stirling cycle

The Stirling cycle cryocooler is among the earliest and most widely adopted types in practice, known for its high efficiency and compact design. As of 2019, more than 200 000 units had been produced worldwide. Achieving high performance with this architecture requires careful engineering of moving pistons, regenerator matrices, and tight thermal control. Table 3.1 summarises its main advantages and disadvantages.

The Stirling cycle resembles the Carnot cycle in its overall temperature span and reversibility but replaces the isentropic processes with regenerative isochoric heat exchange. This substitution allows for more feasible pressure and volume ratios in real machines while preserving high efficiency.

The four process steps of the Stirling cycle, illustrated in figure 3.2, are as follows.

1. **Isothermal compression (a to b):** The compression piston moves inward while maintaining constant temperature T_H, with heat being rejected to the surroundings.
2. **Isochoric heat transfer (b to c):** The gas flows through the regenerator at constant volume, depositing thermal energy into the regenerator matrix.
3. **Isothermal expansion (c to d):** The expansion piston withdraws, allowing the gas to absorb heat from the cooled object at constant temperature T_C, providing useful refrigeration.
4. **Isochoric heat recovery (d to a):** The gas flows back through the regenerator and recovers the previously stored heat, completing the cycle.

The isochoric regeneration steps (b–c and d–a) are essential to achieving high thermal efficiency. As the regenerator absorbs and re-releases the same amount of heat cyclically, there is no net heat transfer across it, and it contributes minimal entropy generation if well designed. This regenerative action allows the Stirling cycle to approach the Carnot limit under idealised, reversible conditions.

The working gas is typically helium due to its low viscosity, high thermal conductivity, and favorable thermodynamic properties. Regenerators are constructed using fine metal mesh or porous structures with high heat capacity and low flow resistance to maximise effectiveness.

Table 3.1. Advantages and disadvantages of Stirling cryocoolers.

Advantages	Disadvantages
High efficiency	Oil-free operation required (esp. at cold end)
Moderate cost	Intrinsic vibration from displacer
Small size and weight	Long lifetime so expensive

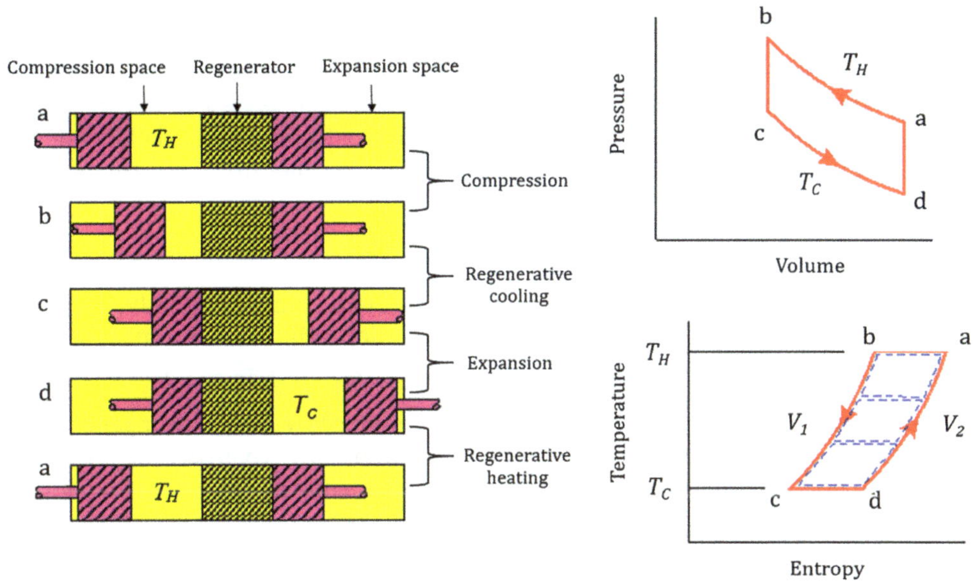

Figure 3.2. Left: A Stirling cycle cooler system showing piston movement and the regenerator. Right: pressure–volume (*P*–*V*) and temperature–entropy (*T*–*S*) diagrams illustrating the four steps of the ideal Stirling cycle. (Adapted from Radebaugh (2000). Image credit: Ray Radebaugh.)

The COP for an ideal reversible Stirling cycle matches that of the Carnot cycle,

$$\mathrm{COP} = \frac{T_C}{T_H - T_C}, \tag{3.64}$$

as all heat transfer processes occur isothermally and all compression/expansion is internally regenerated. However, in practical implementations, pressure drops, regenerator inefficiencies, imperfect sealing, and finite heat exchange reduce the actual COP significantly. Nevertheless, Stirling cryocoolers remain one of the most efficient and scalable options for space and terrestrial cryogenic applications.

3.4.2.2 Gifford–McMahon cycle

The GM cycle is widely used in commercial and laboratory cryocoolers. It features discrete valve timing to control flow direction and timing between the compressor and the expansion space. The rotary valve alternates high/low pressure; the cold space undergoes near-adiabatic expansion/compression, with heat exchange through the regenerator. No throttling (JT) valve is used in the ideal GM process. The use of check valves and phase-specific opening/closing mechanisms allows simplified mechanical implementation but results in pulsating flow and significant pressure loss, limiting efficiency compared to Stirling designs.

3.4.2.3 Pulse tube cycle

Pulse tube cryocoolers eliminate all moving parts at the cold end, enhancing reliability and minimising vibration, an essential feature for sensitive detectors and spaceborne platforms. Oscillatory flows are created by an acoustic driver (such as a piston or pressure wave generator) and shaped by inertance tubes, orifices, and buffer volumes to induce the necessary phasing between pressure and mass flow. While the absence of a displacer reduces mechanical complexity, achieving high performance depends critically on phase control, regenerator performance, and minimising internal loss mechanisms.

3.4.2.4 Joule–Thomson expansion

The JT cryocooler uses isenthalpic expansion through a throttling valve or porous plug to achieve cooling. The cooling capacity depends on the Joule–Thomson coefficient $\mu_{\mathrm{JT}} = \left(\frac{\partial T}{\partial p}\right)_{\mathrm{h}}$, which varies with gas type and initial state. Effective operation often requires precooling stages and counterflow heat exchangers to bring the gas to an appropriate temperature and pressure before expansion. JT coolers are commonly used for final cooling stages in hybrid systems, including achieving sub-4 K temperatures in space applications.

3.4.3 Cycle integration and staging

Many advanced cryocoolers use hybrid or cascaded cycles to enhance performance. For example, a JT stage may be appended to a pulse tube precooler to achieve sub-4 K temperatures. In such configurations, entropy minimisation in one stage supports improved exergy utilisation in the next. The total system COP becomes

$$\mathrm{COP}_{\mathrm{total}} = \frac{\dot{Q}_{\mathrm{c}}}{\dot{W}_{\mathrm{total}}}, \tag{3.65}$$

with $\dot{W}_{\mathrm{total}}$ summing the input powers of all stages.

Cycle matching, interstage heat exchange, and flow impedance management are crucial to maximising integrated performance.

3.4.4 Cycle analysis using exergy

Exergy analysis provides a consistent framework for comparing the effectiveness of each thermodynamic cycle. For any cycle

$$\dot{W}_{\mathrm{rev}} = \dot{Q}_{\mathrm{c}}\left(\frac{T_0 - T_{\mathrm{c}}}{T_{\mathrm{c}}}\right), \tag{3.66}$$

gives the reversible work requirement. Real power input $\dot{W}$ exceeds this by

$$\dot{W} = \dot{W}_{\mathrm{rev}} + T_0\dot{S}_{\mathrm{irr}}, \tag{3.67}$$

and the second-law efficiency is

$$\eta_{\mathrm{II}} = \frac{\dot{W}_{\mathrm{rev}}}{\dot{W}}. \tag{3.68}$$

This framework allows direct comparison between different cryocooler types and stages, accounting for losses due to entropy generation and guiding optimisation toward cycles with superior exergy retention.

In the next section, we delve into pressure–volume (*P–V*) diagrams to illustrate these cycles geometrically and relate the area enclosed in the diagram to work done over a thermodynamic loop.

3.5 Pressure–volume behaviour

P–V diagrams are a powerful tool for visualising thermodynamic cycles, in particular in systems such as cryocoolers that involve compression, expansion, and internal heat exchange processes. These diagrams plot the instantaneous pressure against volume for a fixed mass of working gas as it evolves through a cyclic process. Each closed loop on the diagram represents a thermodynamic cycle, with the area enclosed corresponding to the net work done per cycle.

3.5.1 Physical interpretation

In the context of cryocoolers, the *P–V* diagram offers intuitive insights into the energy exchange mechanisms occurring within the working gas. During compression, the system moves along a curve toward higher pressure and lower volume, and during expansion it moves back toward lower pressure and higher volume. The nature of these paths, whether they are isothermal, adiabatic, or otherwise, determines the shape of the loop.

For example, in an ideal Stirling cryocooler, the *P–V* diagram consists of two isothermal processes (compression and expansion) connected by two constant-volume regenerative processes. The compression and expansion strokes form smooth curves, while the regeneration steps appear as vertical lines since the volume remains constant while pressure changes.

A representative example of the *P–V* diagram, along with the corresponding temperature–entropy (*T–S*) diagram, was shown earlier in figure 3.2 in the context of the Stirling cycle. These diagrams illustrate how thermodynamic trajectories map to piston motion and internal heat exchange stages, reinforcing the connection between physical processes and idealised cycle models.

3.5.2 Work and energy calculations

The area enclosed by the cycle on a *P–V* diagram quantifies the net work done by the system per cycle:

$$W_{\text{cycle}} = \oint p \, dV. \tag{3.69}$$

For a refrigeration system, this work input is required to extract heat from a lower temperature reservoir. The shape and orientation of the loop indicate whether the system performs net work (as in a heat engine) or consumes it (as in a refrigerator). For cryocoolers, the loop typically proceeds in a counter-clockwise direction, indicating that external work is being done on the gas to achieve refrigeration.

3.5.3 Conclusion

P–V diagrams serve not only as a diagnostic and illustrative tool but also as a gateway to more rigorous energy and entropy analyses. Understanding the shape, orientation, and enclosed area of these diagrams allows for direct insights into system efficiency, irreversibility, and component performance. Throughout this book, we will return to *P–V* behaviour in different contexts, using both analytical and numerical tools to quantify system dynamics.

References

Radebaugh R 2000 Development of the pulse tube refrigerator as an efficient and reliable cryocooler *Proc. Inst. Refrigeration* **96** 11–29

Radebaugh R 2003 Thermodynamics of regenerative refrigerators *Generation of Low Temperature and Its Applications* ed T Ohtsuka and Y Ishizaki (Kamakura: Shonan Technology Center) pp 1–20

Walker G 1983 *Cryocoolers. Part 1* (New York: Springer)

IOP Publishing

Mathematical Methods for Cryocoolers

Hannah Rana

Chapter 4

Harmonic approximations

4.1 Harmonic oscillators

Harmonic oscillators form the mathematical foundation for many dynamic processes encountered in cryocooler systems. From piston-driven compressors to oscillatory flow in regenerators and pulse tubes, the behaviour of key subsystems can be approximated as simple harmonic or damped harmonic oscillations under certain operating regimes. This section introduces the governing equations, solutions, and physical interpretations of such systems, laying the groundwork for phasor-based modelling and coupled-mode analysis in later sections.

4.1.1 The simple harmonic oscillator

The simplest form of oscillatory behaviour arises in a system where a restoring force is directly proportional to displacement. This principle underpins a vast class of mechanical and acoustic systems in cryocooler design. Consider a mass m attached to an ideal linear spring with stiffness k, moving without friction on a single axis. According to Newton's second law, the net force acting on the mass is equal to the product of its mass and acceleration. Simultaneously, Hooke's law tells us that the restoring force exerted by the spring is $-kx$, acting to return the mass to equilibrium. Combining these, the equation of motion becomes

$$m\ddot{x}(t) + kx(t) = 0, \tag{4.1}$$

where $x(t)$ denotes the instantaneous displacement from equilibrium and $\ddot{x}(t)$ is the corresponding acceleration. This second-order linear differential equation defines a *simple harmonic oscillator* (SHO), whose dynamics are governed entirely by the system's mass and stiffness.

The general solution to equation (4.1) is given by

$$x(t) = A\cos(\omega_0 t) + B\sin(\omega_0 t), \tag{4.2}$$

doi:10.1088/978-0-7503-4826-3ch4

where A and B are real constants determined by the initial displacement and velocity, and $\omega_0 = \sqrt{k/m}$ is the natural angular frequency of the oscillator. The angular frequency defines the rate at which the system oscillates about the equilibrium point and is a function solely of its inertial and elastic properties. Physically, this implies that lighter masses or stiffer springs produce faster oscillations, while heavier or more compliant systems oscillate more slowly.

In cryogenic engineering, the SHO serves as a powerful abstraction for understanding time-varying behaviours in components where oscillatory motion or flow is intrinsic to operation. One common example is the flexure-suspended piston in a Stirling compressor. These pistons are typically mounted on flexure springs that exhibit approximately linear restoring behaviour over small displacements. During compressor operation, the piston undergoes sinusoidal oscillations about its mean position, and the dynamic response can often be captured using the SHO model, particularly when losses are small and drive amplitudes are modest (Rana 2022a).

Another instance where SHO modelling is applicable arises in pressure wave propagation within cryocooler cavities. In particular, the standing wave patterns that form in certain pulse tube geometries, especially those with inertance tubes and buffer volumes, can exhibit node–antinode behaviour analogous to mass–spring systems. For example, a gas parcel oscillating between two regions of compressibility may be treated as a lumped mass attached to an elastic 'spring' representing gas compliance (Rana 2022b). In this way, spatially extended acoustic fields can be discretised into SHO-like elements that approximate their temporal evolution.

Furthermore, SHO analysis plays an important role in the design and characterisation of acoustic resonators, which form the basis of many thermoacoustic and pulse tube cryocoolers. When a cavity is driven near its natural frequency, small perturbations can give rise to large amplitude oscillations, an effect known as resonance, which is inherently described by SHO dynamics. This underlines the importance of accurate frequency matching between driving and resonant elements for efficient energy transfer and optimal cooling performance.

Although the SHO model is highly idealised, it provides an essential first step in understanding and analysing oscillatory systems. It forms the foundation upon which more complex dynamics, such as damping, external forcing, and coupling between multiple degrees of freedom, can be systematically built. These additional effects are addressed in subsequent sections and are critical for capturing the full behaviour of practical cryocoolers operating under real-world conditions.

4.1.2 Energy in harmonic oscillators

In an SHO, the mechanical energy of the system continuously oscillates between kinetic and potential forms This cyclical energy exchange lies at the heart of dynamic processes in cryocoolers, where the working gas or mechanical components store and release energy as they oscillate in time.

The total mechanical energy E of an undamped SHO is the sum of the instantaneous kinetic energy T and potential energy V:

$$E = T + V = \frac{1}{2}m\dot{x}^2 + \frac{1}{2}kx^2. \tag{4.3}$$

Since there are no dissipative forces, this energy remains constant over time. However, the partitioning between kinetic and potential energy varies throughout the cycle.

At maximum displacement $x = \pm A$, where A is the amplitude, the velocity is zero, and all energy resides in the spring as potential energy:

$$E = V_{\max} = \frac{1}{2}kA^2. \tag{4.4}$$

Conversely, at the equilibrium position $x = 0$, the spring is momentarily relaxed, and all energy is kinetic:

$$E = T_{\max} = \frac{1}{2}m\dot{x}_{\max}^2 = \frac{1}{2}m\omega_0^2A^2, \tag{4.5}$$

where $\omega_0 = \sqrt{k/m}$ is the natural frequency of the system.

To visualise this transformation, we can express the energies as time-dependent functions. If the solution to the SHO is

$$x(t) = A\cos(\omega_0 t + \phi), \tag{4.6}$$

then the velocity is

$$\dot{x}(t) = -A\omega_0 \sin(\omega_0 t + \phi). \tag{4.7}$$

Substituting into the energy expressions, we obtain

$$T(t) = \frac{1}{2}m\omega_0^2A^2\sin^2(\omega_0 t + \phi), \tag{4.8}$$

$$V(t) = \frac{1}{2}kA^2\cos^2(\omega_0 t + \phi). \tag{4.9}$$

The total energy remains time-invariant:

$$E = T(t) + V(t) = \frac{1}{2}kA^2 = \text{constant}. \tag{4.10}$$

This sinusoidal exchange between energy forms, oscillating at frequency $2\omega_0$, illustrates the SHO's conservative nature.

In the context of cryocoolers, this energy oscillation mirrors the enthalpy exchange processes occurring in both the mechanical and gas-dynamic domains. For instance, in a Stirling compressor, the moving piston compresses the gas, storing energy in the form of pressure–volume work, which is then partially converted back into piston motion during the expansion phase. Similarly, in the acoustic core of a pulse tube cryocooler, oscillating gas parcels alternately compress and accelerate through the system, shuttling energy between potential (pressure-related) and kinetic (velocity-related) forms. Although thermal effects are superimposed on these dynamics, the underlying energy exchange mechanism remains fundamentally oscillatory.

Another useful quantity to define is the *energy density* associated with harmonic motion. If the oscillating mass is distributed along a one-dimensional continuum (e.g. a gas column), the local energy density per unit length is given by

$$\mathcal{E}(x, t) = \frac{1}{2}\rho u(x, t)^2 + \frac{1}{2}\kappa p(x, t)^2, \tag{4.11}$$

where $u(x, t)$ is the local particle velocity, $p(x, t)$ the local pressure perturbation, ρ the mean gas density, and κ a compliance-like coefficient related to the inverse of bulk modulus. This form is particularly important in distributed-parameter models of thermoacoustic engines, where standing or travelling waves carry energy across the system without net mass transport.

In cryogenic applications, careful tuning of system parameters ensures that this energy oscillation leads to useful work extraction or effective enthalpy pumping. When phasing and impedance matching are correct, the system achieves a net energy flow from the compression zone to the cold tip, a process made possible by the coherent interplay of kinetic and potential energy oscillations.

The study of energy in harmonic systems, while elementary in isolation, thus provides a conceptual bridge between mechanical dynamics and thermodynamic transport in oscillating cryocoolers. It also sets the stage for evaluating the effects of damping, external forcing, and modal interactions, which will be explored in the sections that follow.

Figure 4.1 illustrates the temporal evolution of kinetic energy $T(t)$, potential energy $V(t)$, and total mechanical energy E in an undamped simple harmonic oscillator over one complete cycle. As shown, $T(t)$ and $V(t)$ oscillate sinusoidally and are perfectly out of phase, such that when one reaches a maximum, the other is

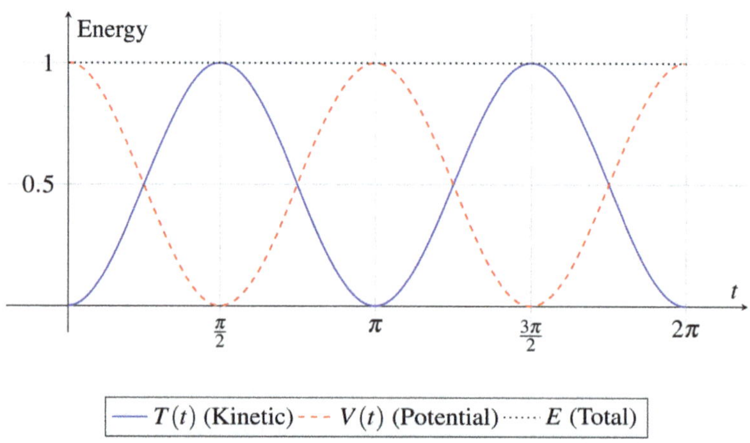

Figure 4.1. Energy exchange in a simple harmonic oscillator over one cycle. Kinetic energy $T(t)$ and potential energy $V(t)$ oscillate out of phase, while the total mechanical energy E remains constant.

minimised. At times $t = 0, \pi, 2\pi$, the system is at maximum displacement and hence stores all energy as potential energy, while the kinetic energy vanishes. Conversely, at $t = \pi/2, 3\pi/2$, the system passes through the equilibrium position with maximum speed, and the energy is entirely kinetic. The total energy E, represented by the horizontal dotted line, remains constant throughout the motion, reaffirming the conservative nature of the SHO. This cyclical transfer between energy forms underpins many processes in cryocooler operation, where pressure-induced compression and expansion phases produce analogous exchanges between kinetic and potential energy in the gas and mechanical subsystems.

4.1.3 Damped and driven oscillations

In real cryogenic systems, idealised oscillatory motion is modified by the presence of energy dissipation mechanisms such as mechanical friction, gas drag, and electromagnetic losses. These damping effects, although typically small, influence the stability, phase response, and amplitude of motion in key cryocooler components such as the compressor piston and displacer.

To account for such losses, a linear damping term proportional to velocity is added to the equation of motion. This yields the standard form of the damped harmonic oscillator,

$$m\ddot{x}(t) + c\dot{x}(t) + kx(t) = 0, \tag{4.12}$$

where c is the damping coefficient. The solution behaviour depends on the damping ratio

$$\zeta = \frac{c}{2\sqrt{km}}, \tag{4.13}$$

which characterises the system's response regime:

- For $\zeta < 1$, the system is *underdamped*, exhibiting oscillatory decay.
- For $\zeta = 1$, the system is *critically damped*, returning to equilibrium as fast as possible without oscillation.
- For $\zeta > 1$, the system is *overdamped*, returning to equilibrium slowly and without overshoot.

Cryocooler compressors are often modelled as underdamped driven systems, particularly when operated near their design resonance frequency. For instance, in a linear compressor, the piston assembly (of mass m) is mounted on flexures (effective stiffness k) and driven by an electromagnetic force generated by alternating current in the coil. The damping term c captures losses from internal friction, eddy currents, and gas back-pressure. Ensuring that $\zeta \ll 1$ allows for high oscillation amplitude with minimal input power, which is essential for efficient thermodynamic cycling.

Displacers in Stirling cryocoolers are similarly modelled using damped oscillators, often treated as pneumatically coupled masses subject to gas spring forces and pressure differentials. While they are not actively driven in classical Stirling configurations, their motion is entrained by pressure oscillations in the system.

However, in free-piston and pulse tube variants, active or semi-active control of displacer dynamics necessitates precise modelling of damping and phasing.

When an external time-periodic force is applied, such as a sinusoidal electromagnetic force in a linear motor, the equation becomes that of a *driven damped oscillator*:

$$m\ddot{x}(t) + c\dot{x}(t) + kx(t) = F_0 \cos(\omega t), \tag{4.14}$$

where F_0 is the amplitude of the driving force and ω is the driving angular frequency. After initial transients decay, the system reaches a steady-state solution of the form

$$x_{ss}(t) = X(\omega)\cos(\omega t - \phi), \tag{4.15}$$

where the amplitude and phase lag are functions of the system parameters and the drive frequency:

$$X(\omega) = \frac{F_0/m}{\sqrt{(\omega_0^2 - \omega^2)^2 + (2\zeta\omega_0\omega)^2}}, \tag{4.16}$$

$$\phi(\omega) = \tan^{-1}\left(\frac{2\zeta\omega_0\omega}{\omega_0^2 - \omega^2}\right). \tag{4.17}$$

These expressions are critical in cryocooler compressor design. The peak amplitude occurs at the *resonant frequency* ω_r, which, in the presence of damping, is slightly lower than the undamped natural frequency ω_0. Specifically, the resonance condition is

$$\omega_r = \omega_0\sqrt{1 - 2\zeta^2}. \tag{4.18}$$

For small damping ($\zeta \ll 1$), this reduction is negligible, but for systems with significant viscous or electromagnetic losses, accurate tuning of ω to ω_r becomes essential for maintaining stroke amplitude without excessive input power.

In many cryocooler systems, phase alignment between the compressor piston and the pressure oscillation it generates is also crucial. From the expression for $\phi(\omega)$, we see that at frequencies below resonance, the displacement lags the drive force only slightly, whereas near resonance, a substantial phase lag emerges. This phase lag affects the timing of compression and expansion in the working gas, directly impacting the thermodynamic efficiency of the cycle.

A similar analysis applies to displacer motion. Although displacers are not typically driven directly, their dynamics can still be captured by a driven response to time-varying pressure differences across the regenerator and cold space. The effective drive force in this case is proportional to $\Delta p(t)$, and the displacer's response will exhibit a frequency-dependent amplitude and phase shift, both of which must be matched carefully to the cycle timing for effective heat exchange.

In summary, modelling the compressor and displacer dynamics using damped and driven harmonic oscillator theory provides both a predictive and diagnostic framework. It enables cryocooler designers to select system parameters (mass, stiffness, damping, and drive frequency) that maximise mechanical efficiency, ensure stable oscillation, and align critical phases of motion with thermodynamic transitions. These principles will later be extended using phasor analysis and impedance methods in section 4.3.

4.1.3.1 Frequency response and resonance behaviour

When a harmonic oscillator is subjected to a periodic external force, its response amplitude depends strongly on the driving frequency. This dependence is captured by the steady-state solution to the driven damped oscillator equation:

$$m\ddot{x} + c\dot{x} + kx = F_0 \cos(\omega t), \tag{4.19}$$

where F_0 and ω denote the amplitude and frequency of the driving force, respectively. The resulting displacement oscillates at the same frequency as the drive, but with an amplitude $X(\omega)$ and a phase lag $\phi(\omega)$ that vary with frequency.

The amplitude response is given by

$$X(\omega) = \frac{F_0/m}{\sqrt{(\omega_0^2 - \omega^2)^2 + (2\zeta\omega_0\omega)^2}}, \tag{4.20}$$

where $\omega_0 = \sqrt{k/m}$ is the natural frequency and $\zeta = c/(2\sqrt{km})$ is the damping ratio. Figure 4.2 shows the normalised amplitude response as a function of normalised driving frequency ω/ω_0, for several representative values of ζ.

As shown, the response exhibits a pronounced resonance peak near $\omega = \omega_0$ in the underdamped regime ($\zeta < 1$). For low damping, the amplitude can grow

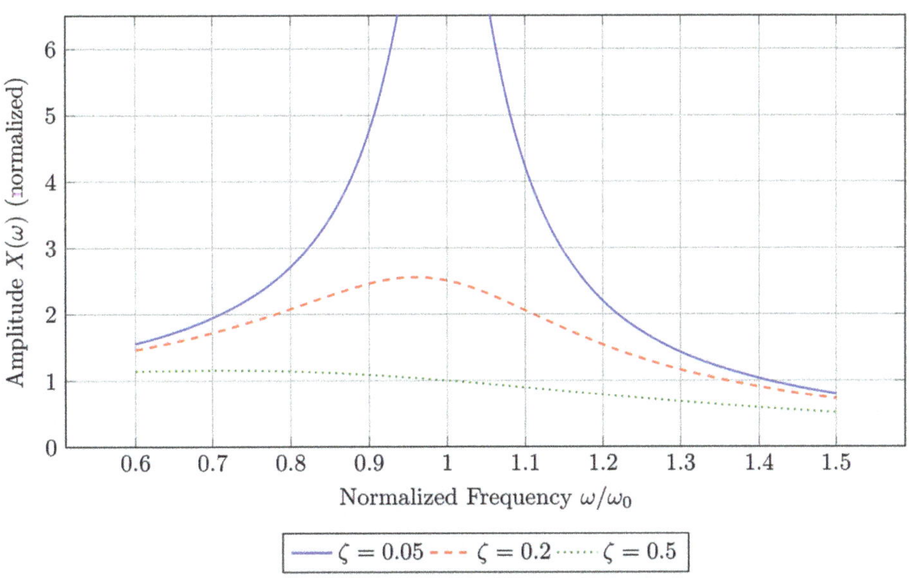

Figure 4.2. Frequency response of a driven damped harmonic oscillator. Lower damping results in higher and sharper resonance peaks. Amplitude is normalised by F_0/m and frequency by the natural frequency ω_0.

significantly at resonance, with the peak becoming sharper and more pronounced as ζ decreases. Conversely, in more heavily damped systems, the resonance is flattened and shifted, reducing both the peak amplitude and the system's sensitivity to drive frequency.

This behaviour has direct implications for cryocooler performance. In linear compressors, for instance, the piston assembly is driven near resonance to maximise stroke amplitude and minimise input power for a given force. A well-tuned system ensures that the operating frequency lies close to the mechanical resonance, while avoiding excessive vibration and mechanical stress. Similarly, in pulse tube and free-piston Stirling systems, the dynamic response of the displacer or acoustic components must be aligned with the thermodynamic cycle timing, which is strongly influenced by the underlying frequency response curve.

The amplitude profile also informs design trade-offs: while low damping improves efficiency by concentrating energy near resonance, it also increases susceptibility to frequency drift and mechanical instability. Cryocooler designers must therefore carefully balance damping, stiffness, and drive frequency to optimise both energy transfer and system robustness.

4.1.4 Resonance and quality factor

A key characteristic of driven harmonic systems is *resonance*: the sharp amplification of oscillatory response when the driving frequency ω closely matches the system's natural frequency ω_0. At resonance, the energy supplied by the external force constructively interferes with the system's motion, leading to a significant build-up in amplitude—especially when damping is low. This phenomenon is particularly pronounced in systems where the damping ratio ζ is small, resulting in sustained high-amplitude oscillations.

The *quality factor* Q quantifies the sharpness and selectivity of the resonance peak. It is defined as the ratio of stored to dissipated energy per cycle and can be expressed in terms of the damping ratio as

$$Q = \frac{\omega_0}{2\zeta\omega_0} = \frac{1}{2\zeta}. \tag{4.21}$$

This shows that as damping decreases, Q increases, implying narrower and more pronounced resonance peaks. High-Q systems exhibit strong sensitivity to driving frequency and are capable of oscillating for many cycles before dissipating their stored energy.

In cryocooler applications, the quality factor plays a central role in determining system efficiency and dynamic behaviour. For example, in Stirling-type compressors, the oscillating piston mass and spring-like restoring forces from flexure bearings form an effective resonant system. Matching the compressor's mechanical resonance to the acoustic resonance of the gas column within the cold head maximises power transfer efficiency. If the mechanical quality factor is too low, damping will dissipate energy prematurely, reducing the pressure amplitude and degrading overall cooling performance.

Similarly, the displacer in a Stirling or pulse tube cooler acts as a resonant mass subjected to oscillatory gas forces. Its motion must remain phase-locked with the pressure oscillations to maintain constructive gas flow through the regenerator. A high-quality factor ensures that the displacer maintains a strong, stable amplitude with minimal external drive force.

The bandwidth of resonance is also related to Q and is often defined in terms of the *full width at half maximum* (FWHM) $\Delta\omega$ of the frequency response curve:

$$\Delta\omega = \frac{\omega_0}{Q}. \tag{4.22}$$

This highlights the tradeoff between frequency selectivity and damping: narrow-band operation improves phase coherence but becomes increasingly sensitive to parameter drift or thermal variations, which are common in space-based cryocoolers.

Finally, the amplitude response function for the driven damped oscillator at frequency ω can be rewritten explicitly in terms of Q as

$$X(\omega) = \frac{F_0/m}{\omega_0^2\sqrt{(1 - (\omega/\omega_0)^2)^2 + \left(\frac{1}{Q} \cdot \frac{\omega}{\omega_0}\right)^2}}. \tag{4.23}$$

This formulation underscores the centrality of Q in shaping the dynamic frequency response and provides a useful basis for characterising compressor impedance, drive electronics, and cold head design in high-performance cryogenic systems.

4.1.5 Applications to cryocoolers

In cryogenic systems, oscillatory dynamics play a foundational role in the operation of key subsystems. The motion of pistons and displacers in Stirling cryocoolers, for instance, is governed by mass–spring dynamics and can be effectively modelled as damped harmonic oscillators. Similarly, the pressure-driven oscillatory flow of gas within pulse tubes and inertance tubes exhibits frequency-dependent behaviour that reflects both resonance and phase lag effects. Even the regenerator, a porous medium central to energy exchange, exhibits resonant thermoacoustic interactions as the gas shuttles back and forth under oscillating pressure gradients.

These elements can often be idealised as individual or coupled harmonic oscillators, allowing for tractable modelling of their dynamic behaviour. Understanding the amplitude and phase response of each component enables designers to match acoustic impedances, minimise dissipative losses, and tune system frequencies to maximise performance. In particular, analysing damping and resonance provides direct insight into how to optimise the geometry, mass distribution, and stiffness of mechanical components, as well as the lengths and diameters of fluidic channels, to achieve efficient and stable cryocooler operation.

4.2 Normal modes

While single-degree-of-freedom harmonic oscillators are foundational, many components in cryocoolers exhibit coupled dynamics. For example, in a Stirling cryocooler, the motion of the displacer is coupled to that of the compressor via the oscillating gas column. Similarly, in pulse tube cryocoolers, pressure and flow oscillations in different parts of the system are interdependent. The framework of *normal modes* provides a systematic way to analyse these multi-degree-of-freedom systems.

4.2.1 Coupled harmonic systems

Consider a simple coupled system consisting of two masses, m_1 and m_2, connected by linear springs of stiffness k as shown schematically in figure 4.3. Let $x_1(t)$ and $x_2(t)$ denote the displacements of the two masses from equilibrium. The Newtonian equations of motion, neglecting damping, are

$$m_1\ddot{x}_1 = -k(x_1 - x_2), \tag{4.24}$$

$$m_2\ddot{x}_2 = -k(x_2 - x_1). \tag{4.25}$$

These equations form a system of coupled second-order ordinary differential equations. To solve them, we assume sinusoidal solutions of the form

$$x_j(t) = X_j e^{j\omega t}, \quad j = 1, 2, \tag{4.26}$$

where ω is the angular frequency of oscillation, and X_j are complex amplitudes. Substituting into the equations of motion, we obtain

$$-\omega^2 m_1 X_1 = -k(X_1 - X_2), \tag{4.27}$$

$$-\omega^2 m_2 X_2 = -k(X_2 - X_1). \tag{4.28}$$

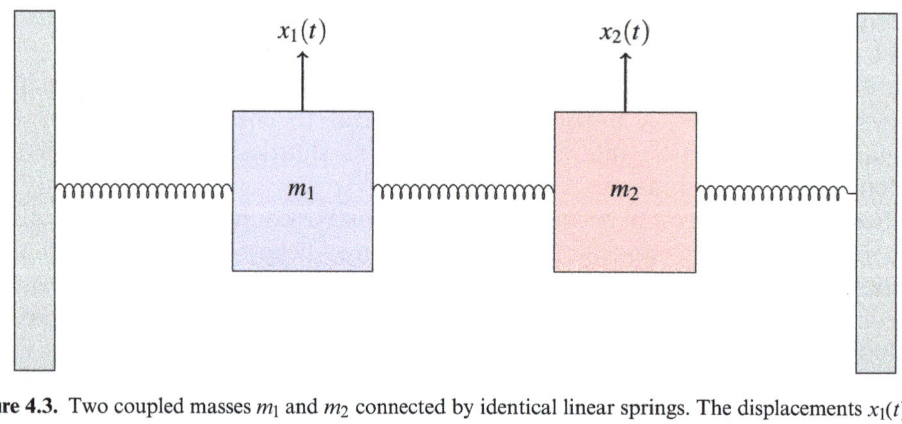

Figure 4.3. Two coupled masses m_1 and m_2 connected by identical linear springs. The displacements $x_1(t)$ and $x_2(t)$ are measured from equilibrium.

Rewriting this system in matrix form yields

$$(\mathbf{K} - \omega^2\mathbf{M})\mathbf{X} = \mathbf{0}, \tag{4.29}$$

where $\mathbf{X} = \begin{bmatrix} X_1 \\ X_2 \end{bmatrix}$ is the mode shape vector, and the mass and stiffness matrices are given by

$$\mathbf{M} = \begin{bmatrix} m_1 & 0 \\ 0 & m_2 \end{bmatrix}, \quad \mathbf{K} = k\begin{bmatrix} 1 & -1 \\ -1 & 1 \end{bmatrix}. \tag{4.30}$$

4.2.2 Eigenvalues and normal modes

The condition for nontrivial solutions to exist is that the determinant of the coefficient matrix must vanish:

$$\det(\mathbf{K} - \omega^2\mathbf{M}) = 0. \tag{4.31}$$

This equation defines a polynomial in ω^2, known as the *characteristic equation*. Solving it yields the eigenvalues ω_1^2, ω_2^2, corresponding to the system's squared natural frequencies. Once the eigenvalues are known, the corresponding eigenvectors $\mathbf{X}_1$, $\mathbf{X}_2$ describe the relative motion of the masses at each mode; these are the *normal modes* of the system.

For instance, in the symmetric case where $m_1 = m_2 = m$, the characteristic equation becomes

$$\det\left(k\begin{bmatrix} 1 & -1 \\ -1 & 1 \end{bmatrix} - \omega^2 m\begin{bmatrix} 1 & 0 \\ 0 & 1 \end{bmatrix}\right) = 0, \tag{4.32}$$

which simplifies to

$$\det\begin{bmatrix} k - \omega^2 m & -k \\ -k & k - \omega^2 m \end{bmatrix} = (k - \omega^2 m)^2 - k^2 = 0. \tag{4.33}$$

Solving this yields

$$\omega^2 = 0 \quad \text{or} \quad \omega^2 = \frac{2k}{m}, \tag{4.34}$$

so the two natural frequencies are

$$\omega_1 = 0, \quad \omega_2 = \sqrt{\frac{2k}{m}}. \tag{4.35}$$

The $\omega = 0$ mode corresponds to rigid-body motion (both masses moving in phase), while the higher-frequency mode corresponds to an out-of-phase vibration, with the masses moving symmetrically in opposite directions. These mode shapes define the natural dynamic response of the coupled system.

4.2.3 Relevance to cryocoolers

In cryocoolers, such coupled oscillatory systems arise in the mechanical linkage between the compressor and displacer, in the acoustic interaction between gas columns separated by regenerators or inertance tubes, and in multi-stage systems where thermal or pressure oscillations propagate between stages. The normal mode decomposition helps identify which combinations of oscillatory components are energetically favoured and which are suppressed due to mismatch in mass, stiffness, or damping.

By computing the eigenfrequencies and mode shapes, engineers can tune the system's geometry and material properties to either promote or suppress specific modes, optimise energy transfer, and minimise unwanted resonances that lead to inefficiencies or instability. This modal analysis forms the basis for the design of high-performance pulse tube and Stirling cryocoolers operating at carefully selected frequencies.

4.2.4 Example: compressor–displacer coupling in a Stirling cryocooler

To illustrate the utility of normal mode analysis in cryocooler design, consider a simplified model of a Stirling cryocooler where the compressor piston and displacer are modelled as two coupled oscillators. Let the compressor mass be $m_1 = 0.3$ kg, the displacer mass be $m_2 = 0.2$ kg, and the effective stiffness due to the gas spring between them be $k = 2000$ N m^{-1}. Assume the pistons are constrained to move linearly along a single axis and that damping is negligible for this analysis.

The equations of motion for the system are

$$m_1\ddot{x}_1 = -k(x_1 - x_2), \tag{4.36}$$

$$m_2\ddot{x}_2 = -k(x_2 - x_1), \tag{4.37}$$

which can be written in matrix form as

$$\begin{bmatrix} k & -k \\ -k & k \end{bmatrix}\begin{bmatrix} X_1 \\ X_2 \end{bmatrix} = \omega^2\begin{bmatrix} m_1 & 0 \\ 0 & m_2 \end{bmatrix}\begin{bmatrix} X_1 \\ X_2 \end{bmatrix}.$$

Solving the eigenvalue problem $\det(\mathbf{K} - \omega^2\mathbf{M}) = 0$, we obtain the characteristic equation

$$(k - \omega^2 m_1)(k - \omega^2 m_2) - k^2 = 0.$$

Substituting the numerical values gives

$$(2000 - 0.3\omega^2)(2000 - 0.2\omega^2) - 2000^2 = 0.$$

Expanding and simplifying:

$$(2000^2 - 2000(0.2 + 0.3)\omega^2 + 0.06\omega^4) - 4,000,000 = 0,$$

$$-1000\omega^2 + 0.06\omega^4 = 0 \quad \Rightarrow \quad \omega^2(0.06\omega^2 - 1000) = 0.$$

Solving, we find

$$\omega_1 = 0, \quad \omega_2 = \sqrt{\frac{1000}{0.06}} \approx 129.1 \text{ rad s}^{-1}.$$

The $\omega_1 = 0$ mode corresponds to the compressor and displacer moving together in phase: this is the rigid-body mode. The higher-frequency mode at $\omega_2 \approx 129$ rad s^{-1} represents the out-of-phase oscillation in which energy is exchanged between the compressor and displacer through the intermediate gas spring.

This analysis reveals that tuning the operating frequency of the cryocooler close to the non-zero natural frequency can maximise dynamic coupling between stages. However, if that mode coincides with an undesirable resonance (e.g. excessive vibration in the cold head), the system geometry or mass distribution can be modified to shift ω_2 appropriately. Thus, eigenvalue analysis provides crucial insights into both efficiency and stability of cryogenic system dynamics.

4.2.5 General solution and modal superposition

In a multi-degree-of-freedom (MDOF) system, such as a cryocooler with interacting mechanical and fluidic elements, the motion is governed by a set of coupled second-order differential equations. These equations are typically expressed in the form

$$\mathbf{M}\ddot{\mathbf{x}}(t) + \mathbf{K}\mathbf{x}(t) = \mathbf{0}, \tag{4.38}$$

where $\mathbf{x}(t)$ is the vector of generalised displacements, $\mathbf{M}$ is the mass (or inertance) matrix, and $\mathbf{K}$ is the stiffness (or compliance inverse) matrix. For cryocoolers, the components of $\mathbf{x}(t)$ may correspond to piston positions, displacer motions, or pressure fluctuations at discrete lumped nodes.

The key insight of modal analysis is that the general solution to equation (4.38) can be written as a linear superposition of the system's normal modes. Each normal mode oscillates at a characteristic natural frequency ω_n and has a spatial profile described by an eigenvector $\mathbf{X}_n$:

$$\mathbf{x}(t) = \sum\nolimits_{n=1}^{N} (A_n \cos(\omega_n t) + B_n \sin(\omega_n t))\mathbf{X}_n, \tag{4.39}$$

where N is the number of degrees of freedom, and A_n, B_n are scalar coefficients determined by the system's initial displacement and velocity:

$$\mathbf{x}(0) = \sum\nolimits_{n=1}^{N} A_n \mathbf{X}_n, \tag{4.40}$$

$$\dot{\mathbf{x}}(0) = \sum\nolimits_{n=1}^{N} B_n \omega_n \mathbf{X}_n. \tag{4.41}$$

This decomposition is made possible by the orthogonality properties of the mode shapes $\mathbf{X}_n$. Specifically, the eigenvectors of the generalised eigenvalue problem

$$\left(\mathbf{K} - \omega_n^2 \mathbf{M}\right)\mathbf{X}_n = \mathbf{0}, \tag{4.42}$$

satisfy the following orthogonality relations

$$\mathbf{X}_m^T\mathbf{M}\mathbf{X}_n = 0, \quad \text{for} m \neq n, \tag{4.43}$$

$$\mathbf{X}_m^T\mathbf{K}\mathbf{X}_n = 0, \quad \text{for} m \neq n. \tag{4.44}$$

These relations imply that the modes are decoupled with respect to both inertia and stiffness. Consequently, the full coupled system can be transformed into N independent scalar equations, each representing a simple harmonic oscillator:

$$\ddot{q}_n(t) + \omega_n^2 q_n(t) = 0, \tag{4.45}$$

where $q_n(t)$ is the modal coordinate associated with mode n. This transformation dramatically simplifies both analytical and numerical treatment of cryocooler dynamics, enabling precise modal tuning, fault isolation, and control design.

In practical cryocooler modelling, modal superposition can be used to assess dynamic responses under transient excitation, identify resonant interactions between mechanical and acoustic subsystems, and design feedback strategies to suppress unwanted vibrations in cold-stage assemblies.

Normal mode analysis is particularly useful for understanding dynamic interactions within cryocoolers. Key applications include:

- **Split Stirling cryocoolers**: The displacer and compressor can be modelled as two masses coupled through a gas spring. Modal analysis helps predict piston–displacer interactions and optimise phasing for efficiency.
- **Pulse tube cryocoolers**: Acoustic standing wave modes in the regenerator and pulse tube can be described by normal mode decomposition, particularly when modelled using lumped-parameter analogs.
- **Thermoacoustic engines**: The working gas behaves as a compressible fluid with distributed mass and stiffness. Spatial normal modes determine pressure and velocity field distributions.

4.2.6 Mode tuning and resonant design

Designing cryocoolers for optimal performance often requires *tuning* system parameters to align mode frequencies with desired operating frequencies. For instance, a Stirling cryocooler may be designed such that the displacer resonates near the driving frequency, maximising compression efficiency while maintaining phase control. Alternatively, mode detuning may be employed to suppress unwanted resonances that cause instability or parasitic energy losses.

Advanced systems may even incorporate adjustable inertance tubes or compliance volumes to dynamically shift the modal structure. These strategies will be explored further in section 4.3 and chapter 7.

4.3 Phasor analysis for Stirling and pulse tube cryocoolers

Oscillating cryocoolers such as Stirling and pulse tube systems operate with time-varying pressure, flow, and displacement fields that are near-sinusoidal under steady conditions. Representing these variables as phasors, complex-valued quantities at a fixed driving frequency, enables an elegant formulation of energy transfer and impedance interactions in the frequency domain.

4.3.1 Mechanical and acoustic impedance

Cryocooler subsystems can often be modelled as networks of reactive and resistive components analogous to electrical circuits. The concept of *mechanical impedance* characterises the opposition to oscillatory motion in mechanical elements and is defined by the complex ratio of force to velocity:

$$\tilde{F} = Z_m \tilde{v}, \qquad Z_m = R + j\omega M - \frac{j}{\omega C}, \tag{4.46}$$

where R represents damping (e.g. mechanical friction), M is an effective inertance (mass-like behaviour), and C is a compliance (inverse stiffness). These quantities encapsulate the dynamic behaviour of components such as pistons, flexures, and suspended displacers.

A closely related concept is *acoustic impedance*, which describes the ratio of complex pressure to volume flow rate in gas systems:

$$\tilde{p} = Z_a \tilde{U}, \tag{4.47}$$

where Z_a captures the aggregate effect of resistive, inertial, and compliant elements in the working gas pathway. For instance, porous regenerators contribute resistance, buffer volumes contribute compliance, and inertance tubes contribute mass-like behaviour. Together, these form a lumped acoustic network that governs phase relations and power transmission through the system.

4.3.2 Phasor models of Stirling cryocoolers

In Stirling cryocoolers, the driving piston imposes a pressure wave that propagates through the regenerator, interacts with the displacer, and reflects from the cold end. Using phasor notation, these relationships can be modelled as

$$\tilde{F}_{\text{piston}} = Z_{\text{mech}} \tilde{v}_{\text{piston}}, \tag{4.48}$$

$$\tilde{p}_{\text{reg}} = Z_{\text{reg}} \tilde{U}_{\text{reg}}, \tag{4.49}$$

$$\tilde{p}_{\text{buffer}} = Z_{\text{buf}} \tilde{U}_{\text{buf}}. \tag{4.50}$$

The phasor representation allows one to formulate the cryocooler as an impedance network, where pressure is analogous to voltage and volume flow rate to current. The system can then be solved using Kirchhoff-like continuity and momentum balance conditions. For example, continuity of flow at junctions and conservation of pressure drop across components leads to a solvable set of complex algebraic equations.

4.3.3 Phasor analysis in pulse tube cryocoolers

Pulse tube cryocoolers benefit even more from phasor techniques due to the absence of moving parts in the cold end. The pressure wave generated by a compressor is partially reflected and partially transmitted through components such as the regenerator, pulse tube, inertance tube, and reservoir. Each of these can be modelled as an impedance element:

$$\tilde{p}_{\text{pt}} = Z_{\text{pt}} \tilde{U}_{\text{pt}}, \tag{4.51}$$

$$\tilde{p}_{\text{inert}} = Z_{\text{inert}} \tilde{U}_{\text{inert}}. \tag{4.52}$$

Phase differences between pressure and flow phasors are critical in achieving a net enthalpy flux in the pulse tube, which forms the basis of cooling. Phasor analysis thus enables direct computation of phase shift, acoustic power flow, and impedance matching, essential tools for optimising pulse tube geometry and operating frequency.

Pulse tube cryocoolers (PTCs) rely on phase differences between pressure and flow at the cold and hot ends of the pulse tube to produce cooling. Phasor analysis enables a quantitative understanding of these phase relationships. A simplified PTC model includes a pressure phasor $\tilde{p}(x)$ that varies along the axis, a flow phasor $\tilde{U}(x)$, and acoustic impedances for the regenerator, pulse tube, inertance tube, and reservoir.

By applying mass continuity and momentum conservation in phasor form, the pressure and flow fields can be computed at each location. For example, the inertance tube impedance is

$$Z_{\text{inertance}} = j\omega\rho\frac{L}{A}, \tag{4.53}$$

where ρ is gas density, L is tube length, and A is cross-sectional area.

The net enthalpy flow $\tilde{H}$ into the cold region is given by

$$\tilde{H} = \frac{1}{2}\Re\{\tilde{p}\,\tilde{U}^*\}, \tag{4.54}$$

where $\tilde{U}^*$ is the complex conjugate of the volume flow rate. Maximum cooling is achieved when pressure and flow are appropriately phased, ideally, close to 90 degrees out of phase at the cold end.

4.3.4 Power and phase considerations

The real power transferred by the working gas can be computed as

$$\bar{P} = \frac{1}{2}\Re\{\tilde{p}\,\tilde{U}^*\}, \tag{4.55}$$

where the phase angle between pressure and flow determines the direction and magnitude of energy transport. Cryocooler performance hinges on correct phase alignment: for example, maximised cooling requires the pressure to lead or lag the flow by an optimal angle, often tuned via the length of inertance tubes or the use of orifice valves.

4.3.5 Worked example: impedance and power flow in a Stirling cryocooler stage

Consider a simplified model of a Stirling cryocooler consisting of a piston, a regenerator, and a buffer volume. At a fixed frequency of $\omega = 2\pi \times 60$ rad s^{-1}, the following parameters are given:

- Mechanical impedance of the piston: $Z_{\text{piston}} = 5 + j3$ Pa s m^{-3}
- Acoustic impedance of the regenerator: $Z_{\text{reg}} = 10 + j8$ Pa s m^{-3}
- Complex volume flow rate at piston: $\widetilde{U}_{\text{piston}} = 0.01\, e^{j\frac{\pi}{3}}\ m^3\, s^{-1}$.

Step 1: Compute complex pressure at piston face

$$\widetilde{p}_{\text{piston}} = Z_{\text{piston}}\widetilde{U}_{\text{piston}} = (5 + j3)(0.01e^{j\frac{\pi}{3}}). \tag{4.56}$$

First, convert Z_{piston} to polar form:

$$|Z_{\text{piston}}| = \sqrt{5^2 + 3^2} = \sqrt{34} \approx 5.83, \quad \arg\left(Z_{\text{piston}}\right) = \tan^{-1}\left(\frac{3}{5}\right) \approx 0.5404 \text{ rad}.$$

Thus

$$\widetilde{p}_{\text{piston}} = 5.83 \cdot 0.01 \cdot e^{j\left(\frac{\pi}{3}+0.5404\right)} = 0.0583e^{j1.587} \text{ Pa}.$$

Step 2: Compute average power delivered by the piston

The average acoustic power input to the regenerator is given by

$$\bar{P} = \frac{1}{2}\Re\{\tilde{p}\,\tilde{U}^*\}.$$

Since $\widetilde{p} = P_0 e^{j\theta}$ and $\widetilde{U} = U_0 e^{j\phi}$, their product is

$$\widetilde{p}\,\widetilde{U}^* = P_0\, U_0 e^{j(\theta-\phi)}$$

$$\Rightarrow \Re\{\tilde{p}\,\tilde{U}^*\} = P_0\, U_0 \cos(\theta - \phi).$$

Using

$$P_0 = 0.0583, \quad U_0 = 0.01, \quad \theta - \phi = 1.587 - \frac{\pi}{3} \approx 0.54 \text{ rad}$$

$$\Rightarrow \bar{P} = \frac{1}{2}(0.0583)(0.01)\cos(0.54) \approx \frac{1}{2}(5.83 \times 10^{-4})(0.857) \approx 2.5 \times 10^{-4} \text{ W}.$$

Interpretation: The piston is delivering approximately 0.25 mW of acoustic power into the system. This net energy transfer depends critically on the phase difference between pressure and flow. Had the phase been $\pi/2$, no average power would be delivered, highlighting the importance of impedance matching and phasing in cryocooler design.

4.3.6 Phasor diagrams

4.3.6.1 Pressure and mass flow phase

The key characterising physical components in the pulse tube, pressure and mass flow, can be expressed as a complex number consisting of a complex and real part due to their oscillatory operation in the cryocooler. These can be plotted as phasors on a phasor diagram, as shown in figure 4.4. Given that the pressure in the cryocooler experiences a smaller phase change (5°–10°) compared to the mass flow (anywhere between 50° and 90°), pressure is used as the real axis. Pressure is taken as the reference (real axis), the imaginary axis represents a +90 degree phase shift, and differentiation/integration correspond to multiplying the phasor by $i\,\omega$, or, $1/(i\,\omega)$, respectively. The phasor diagram expressing mass flow at the cold and warm end is shown in figure 4.4.

4.3.6.2 Mass flow at the pulse tube cold end

Taking first the case of an in-line pulse tube cryocooler using an orifice at the warm end where the displacer would be situated, where helium is the working fluid, the pressure pulse $P(t)$ can be written as given in equation (4.57), where P_o is the average pressure, A_p is the amplitude of the pressure wave, and all other terms are as previously defined:

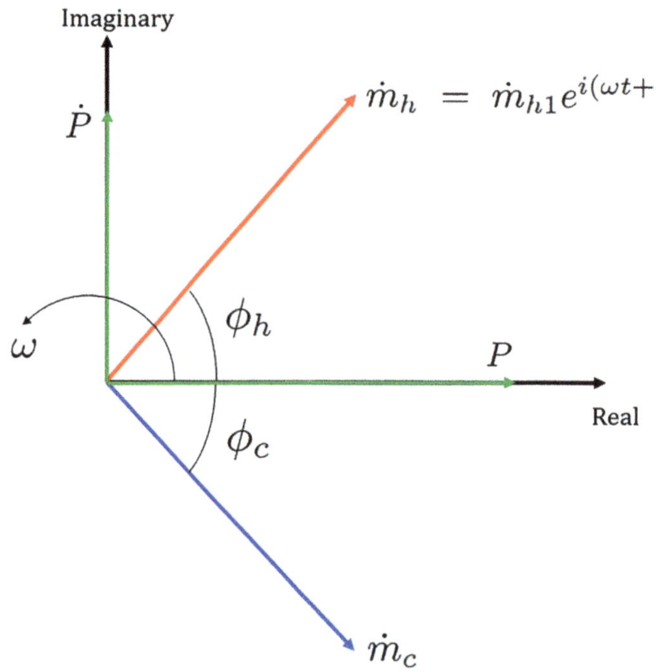

Figure 4.4. Mass flow at the warm end and cold end expressed as phasors against pressure in the real and imaginary axes.

$$P(t) = P_o + A_p \cos(2\pi f t). \tag{4.57}$$

Similarly, the change in mass in the pulse tube can be written as follows, where dm_h and dm_c are the change in mass at hot end and cold end of the pulse tube, respectively (Almtireen 2020):

$$dm_{PT} = dm_h - dm_c. \tag{4.58}$$

The mass flow at the hot end where the orifice is situated is given by

$$\dot{m}_h(t) = C_{or} \cdot \Delta P(t), \tag{4.59}$$

where C_{or} and $\Delta P(t)$ are the flow conductance of the orifice, which is determined as the ratio of mass flow to the pressure drop, and the pressure difference across the orifice, respectively. For an adiabatic process using helium as a monatomic ideal gas, the following thermodynamic relationships hold true and can be utilised (Kittel 1996):

$$P \cdot dV = RT \cdot dm \tag{4.60}$$

$$P \cdot V_{PT}{}^{\gamma} = \text{constant} \tag{4.61}$$

$$\left(\frac{T}{T_0}\right) = \left(\frac{P}{P_0}\right)^{\frac{\gamma-1}{\gamma}} \tag{4.62}$$

$$c_p = \left(\frac{R\,\gamma}{\gamma - 1}\right), \tag{4.63}$$

where R, γ, V_{PT}, and c_p are the ideal gas constant, the ratio of specific heat for helium, the pulse tube volume, and the specific heat capacity of helium for constant pressure. T_0 and P_0 are the average temperature and pressure states. Differentiating equation (4.61) gives

$$V_{PT} \cdot dP + \gamma P \cdot dV_{PT} = 0. \tag{4.64}$$

Substituting this and equation (4.60) into equation (4.58) now allows for an expression for the change in mass flow at the cold end to be obtained:

$$\frac{dm_c(t)}{dt} = \frac{V_{PT}}{\gamma R\, T_c}\frac{dP(t)}{dt} + \frac{T_h}{T_c}\frac{dm_h(t)}{dt}. \tag{4.65}$$

In order to obtain a full relation for the gas mass flow rate, equation (4.59), which provides the mass flow at the hot end as a function of time, can be substituted into equation (4.65). Moreover, a differential form of the pressure pulse can be obtained by differentiating equation (4.57), in order to relate the pressure change terms This gives the following equation for mass flow rate at the cold end:

$$\dot{m}_c(t) = \frac{\omega P V_{PT}}{\gamma R T_c} \cos\left(2\pi f t + \frac{\pi}{2}\right) + \frac{T_h}{T_c} \underbrace{C_{or} P \cos\left(2\pi f t\right)}_{\dot{m}_h}. \tag{4.66}$$

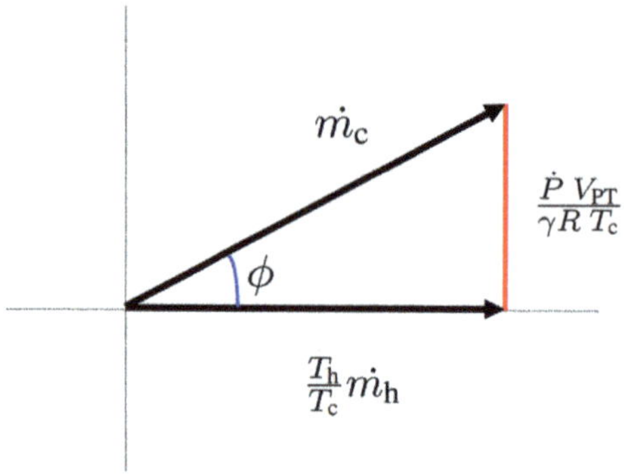

Figure 4.5. Phasor diagram showing the mass flow rates at the cold end and hot end of the pulse tube. The mass flow rate at the cold end is the sum of the two vectors shown and as given by equation (4.66).

From equation (4.66) the sinusoidal nature of the mass flow waveform at the cold head can be seen. It is now further possible to understand how the different parameters hold influence over the mass flow. Notably, the pressure amplitude P and frequency f are variables upon which $\dot{m}_c$ depends, as well as the pulse tube volume V_{PT} and the cold-end temperature T_c. Furthermore, it can be seen that the expression formulates as two vectors, one of which incorporates the expression for $\dot{m}_h$ as shown in equation (4.66). Equation (4.67) expresses this in simplified form, where $\dot{P}$ approximates the angular pressure wave term:

$$\dot{m}_c = \frac{\dot{P}V_{PT}}{\gamma R T_c} + \dot{m}_h\,\frac{T_h}{T_c}. \tag{4.67}$$

Figure 4.5 represents these vectors and their summing to give the mass flow rate at the cold end in the form of a phasor diagram. Hence, the phase angle between the mass flow and the pressure pulse can be formulated accordingly. Equation (4.68) formulates this based on the phasor diagram shown in figure 4.5, where the T_c terms cancel. Interestingly, this now indicates that the mass flow to pressure pulse phase is not influenced by the cold-end temperature:

$$\phi = \tan^{-1}\!\left(\frac{\dot{P}V_{PT}}{T_h\dot{m}_h\gamma R}\right). \tag{4.68}$$

4.3.6.3 SPTC component phasors

Now, for a pulse tube cryocooler using a warm-end displacer (hence, an SPTC), the cold-end mass flow can be expressed following a similar logic with a few modifications. Radebaugh highlights that, given there is no mass flow crossing the moving system boundary but work cross the boundary at that location, an additional expander term is required that accounts for the expander gas volume due to displacer activity (Radebaugh 2003). Integrating equation (4.69), the equation for mass conservation, over the length of the pulse tube, and approximating the pressure pulse to $\dot{P}$ allows equation (4.70) to be expressed (Radebaugh 2003), where A_g is the gas cross-sectional area and T_a is the average temperature in the component:

$$\frac{\partial}{\partial x}\left(\frac{\dot{m}}{A_g}\right) = -\frac{\partial \rho}{\partial t} \tag{4.69}$$

$$\dot{m}_h = \dot{m}_c + \frac{\dot{P}V}{RT_a}. \tag{4.70}$$

Hence, equation (4.71) gives the cold-end mass flow rate for an SPTC using a warm-end displacer, assuming isothermal expansion, in terms of T_c. Given that for SPTCs, comparatively smaller pressure amplitudes are achieved, the average volume during a cycle is taken as half that of the volume for the total swept volume of the displacer (Radebaugh 2003), hence the latter term being halved. As the displacer is situated at the warm, the approximation $T_{disp} \approx T_h$ is made. V_d and V_D denote the instantaneous displacer gas volume and the total displacer swept volume, respectively:

$$\dot{m}_h = \frac{P\dot{V}_d}{RT_h} + \frac{\dot{P}V_D}{2RT_h}. \tag{4.71}$$

Similarly, an equivalent relationship can be formulated for an isothermal compressor, where V_{CO} and V_{co} are the total swept compressor volume and instantaneous compressor gas volume, respectively:

$$\frac{-P\dot{V}_{co}}{RT_{co}} = \dot{m}_{co} + \frac{\dot{P}V_{CO}}{2RT_{co}}. \tag{4.72}$$

Similarly, the mass flow can be expressed for the regenerator in the same manner:

$$\dot{m}_h = \dot{m}_c + \frac{\dot{P}V_{reg}}{RT_{reg}}. \tag{4.73}$$

The regenerator must serve as a very good heat exchanger between the working gas and the regenerator matrix. Across any cross-sectional area at any location within the regenerator, the temperature is assumed as constant, hence $\Delta T_{reg,cs} = 0$. Conversely, a large temperature gradient of $\Delta T = 300\,\text{K}–80\,\text{K}$ occurs across its length. Generally, T_a is taken as the average temperature in the case of the regenerator (Vanapalli 2007).

It can be seen in equation (4.73) that with an increase in frequency or regenerator volume there is a proportional increase in the second term, which forms the vertical phasor vector. As explored in equation (4.82), the mass flow magnitude in the regenerator is directly proportional to regenerator losses. This is due to the fact that with an increase magnitude of $\dot{m}_h$, a larger heat exchanger is required for effective heat transfer, which results in an increased pressure drop and a degraded overall efficiency as a consequence. Second, for a fixed average pressure, an increased $\dot{m}_h$ results in a proportionally increased volumetric flow. This therefore would require a

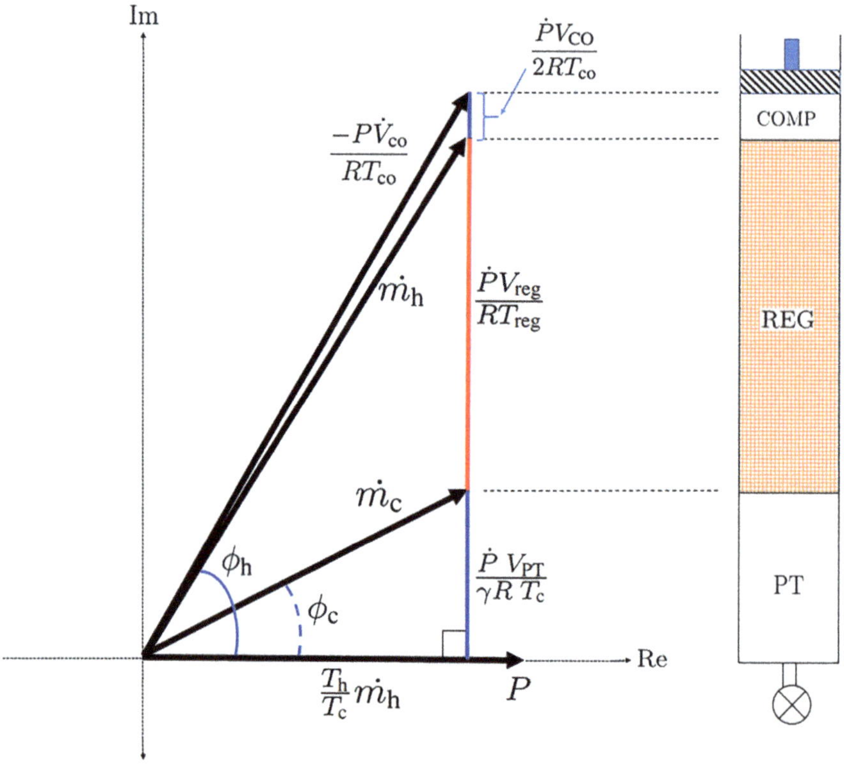

Figure 4.7. Phasor diagram for an orifice pulse tube cryocooler, where equations (4.67), (4.72) and (4.73) are plotted. 'Re' denotes the real axis and 'Im' denotes the imaginary axis. (Image credit: Ray Radebaugh.)

power at the cold end, resulting in a decreased efficiency. This relates to the notion explored that expansion volume is required at the warm end in order to recover PV power and recycle it back into the system, with an optimised phase angle ϕ.

The acoustic power, or PV power, for the pulse tube system is given by

$$\langle \dot{W} \rangle = \langle P\dot{V} \rangle = \frac{1}{2}|P||\dot{V}|\cos\theta = \frac{1}{2}\frac{|P|}{P_0}|\dot{m}|RT\cos\theta \tag{4.80}$$

and, hence,

$$\langle \dot{W} \rangle \propto |\dot{m}|\cos\theta. \tag{4.81}$$

Since acoustic power is measured in the real axis (zeroth derivative), an increase in the amplitude of $\dot{m}$ does not increase PV in the real axis, as per 4.7. This therefore implies that a higher amplitude of $\dot{m}$ results in losses (namely in the regenerator) with no increase in refrigeration power, leading to the relationship of proportionality shown in

$$\text{regenerator losses} \propto |\dot{m}|\cos\theta. \tag{4.82}$$

It is thus favourable to ensure that the amplitude of vertical vector pertaining to the mass flow is minimised. In the phasor diagram for an orifice pulse tube cryocooler shown in figure 4.7, the vertical component of the mass flow vector at the cold end, $\dot{m}_c$, is in the positive imaginary axis, as no warm-end heat exchanger or inertance tube features at the base of the pulse tube to enable $\dot{m}_c$ to lag behind the pressure where the orifice is located. This results in a higher amplitude of the vertical component of the mass flow vector at the warm end, $\dot{m}_h$, in the imaginary axis, leading to higher regenerator losses with no increase in refrigeration power. It is therefore desirable to ensure that the mass flow at the cold end is out of phase such that it lags the pressure pulse, in order to reduce losses in the system. This is why the use of an inertance tube or a displacer to create a phase difference at the other end of the pulse tube is beneficial for achieving higher efficiency in the overall cryocooler system (De Waele 2019).

Furthermore, this sheds light on the results of the comparative study conducted in (Rana 2020), where the in-line pulse tube cryocooler using an inertance tube has been found to achieve a higher cooling power than the in-line SPTC with a warm-end displacer, however, the SPTC yielded a much higher efficiency due to PV recovery at the warm end. This comparative outcome is reflected in the phasor theory that has been discussed in this chapter, where better alignment of the mass flow to pressure phase towards the midpoint of the regenerator results in reduced losses due to minimised mass flow magnitudes and expansion power recovered at the warm end. The phase input into the inertance tube cryocooler numerical model was the optimised phase angle that was experimentally determined for the in-line SPTC system; it is unlikely this phase matched with the optimum ϕ required for the inertance tube pulse tube, in order for the pressure pulse and mass flow to be in phase at the regenerator midpoint. The high value for Q_c and lower value for η_r in the inertance tube simulation suggests the ϕ input was shifted in the positive direction, as depicted for the system shown in figure 4.7. In addition to the under-estimation of conductive and convective losses within the cryocooler, this is a likely explanation for the inertance tube simulated results. Finally, for an in-line SPTC, figure 4.8 depicts the different phasor vectors for the compressor, regenerator, pulse tube, and displacer at the warm end.

4.3.6.6 Phasor analysis of double-inlet cryocoolers

Double-inlet pulse tube cryocoolers (DIPTCs) incorporate an additional orifice and reservoir at the regenerator cold end to manipulate the phasing between pressure and mass flow. This modification improves cooling performance without requiring moving components at the cold head. In sinusoidal steady-state, the cold-end mass flow rate can be written as the sum of a capacitive term and a resistive term:

$$\dot{m}_c = \frac{\dot{P} V_t}{\gamma R T_c} + \left(\frac{T_H}{T_c} \right) \dot{m}_{nt} \tag{4.83}$$

Here, $\dot{P}$ is the pressure time derivative, V_t is the pulse tube volume, T_c and T_H are the cold and hot-end temperatures, respectively, R is the specific gas constant, γ is the heat

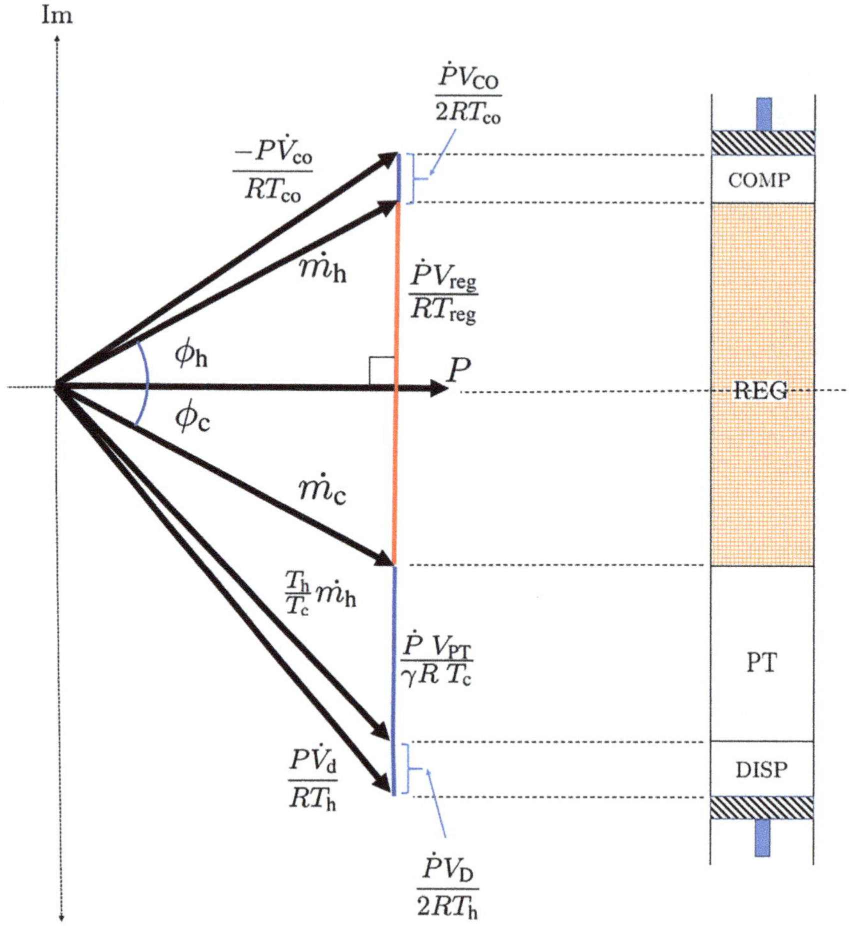

Figure 4.8. Phasor diagram for an in-line SPTC with a warm-end displacer. The cold- and warm-end heat exchangers have been omitted in order to convey the main component phasors based on the formulated expressions equations (4.67)–(4.73).

capacity ratio, and $\dot{m}_{nt}$ is the net mass flow through the warm-end orifice. The first term represents capacitive (imaginary) flow while the second represents resistive (real) flow.

As discussed, this expression for $\dot{m}_c$ is already established in Equation 4.67. Similarly, the expressions for the regenerator mass flow and compressor phasor flow have been derived earlier and are not repeated here. These previous expressions describe how volumetric compression or expansion and void volumes contribute to phase shifts in mass flow.

In the DIPTC configuration, a secondary orifice introduces an additional flow path, governed by

$$\dot{m}_{o2} = C_{o2} \cdot (P_{rg,ce} - P_{res}), \tag{4.84}$$

where C_{o2} is the conductance of the secondary orifice, and $P_{rg,ce}$ and P_{res} are the pressures at the regenerator cold end and reservoir, respectively. This secondary flow

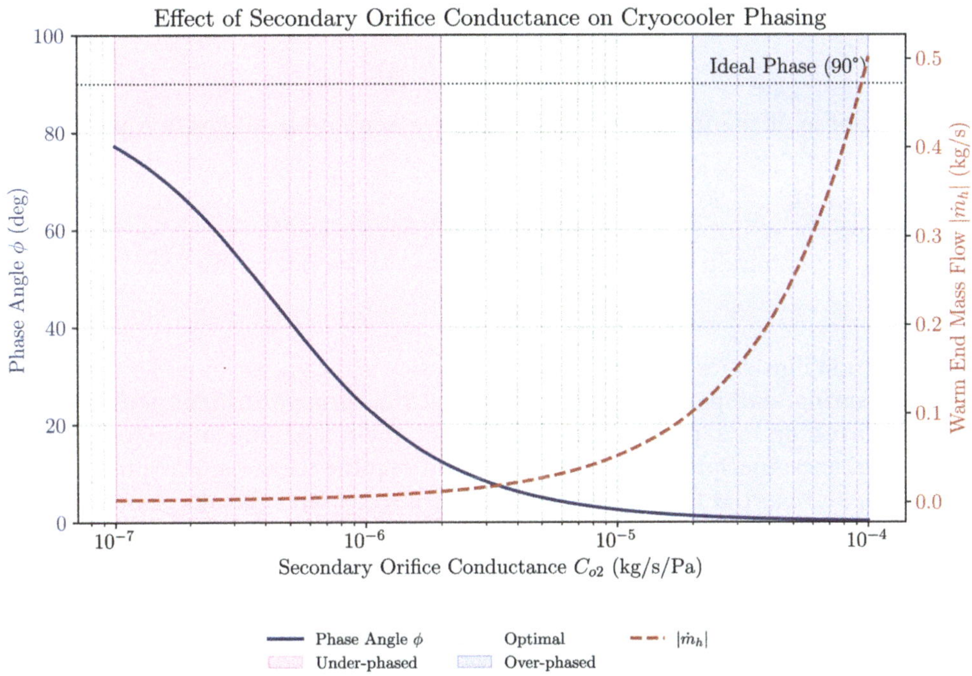

Figure 4.9. Simulation of phase angle ϕ and hot-end mass flow magnitude $|\dot{m}_\mathrm{h}|$ versus secondary orifice conductance. Increased conductance improves phase alignment and reduces regenerator loading.

adjusts the phase angle ϕ between mass flow and pressure at the cold end, reducing the imaginary component of the mass flow vector. As a result, the net flow magnitude is reduced, and the regenerator performance is improved by lowering acoustic losses (Radebaugh 2018, 2019). Figure 4.9 shows simulated results for the variation in ϕ and $|\dot{m}_\mathrm{h}|$ as a function of the secondary orifice conductance. As conductance increases, the phase angle approaches alignment with the pressure waveform and the total phasor magnitude decreases, reflecting more efficient flow phasing and reduced pressure drop. This enables improved enthalpy transfer at the cold end with minimal additional complexity.

As discussed by Radebaugh (2019) and in subsequent numerical studies, double-inlet configurations can approach the performance of active displacer systems, while retaining advantages such as mechanical simplicity and improved long-term reliability.

4.3.6.7 Phasor analysis of the damped harmonic oscillator

Building on the formulation introduced in section 4.1.3, we now recast the damped, driven harmonic oscillator in the frequency domain using phasor analysis. This technique transforms the time-domain differential equation into an algebraic expression, revealing the relationships between forcing, damping, and response more intuitively and enabling efficient analysis of resonant systems such as cryocooler compressors.

Assuming a steady-state sinusoidal forcing term,

$$F_a(t) = F_{a1}\cos(\omega t + \alpha), \tag{4.85}$$

and a response of the form $x(t) = x_1 e^{i\omega t}$, the velocity and acceleration become

$$\dot{x}(t) = i\omega x_1 e^{i\omega t}, \qquad \ddot{x}(t) = -\omega^2 x_1 e^{i\omega t}. \tag{4.86}$$

Substituting these into the driven equation of motion (see equation (4.14)) and cancelling the common exponential factor leads to the phasor-domain form:

$$F_1 e^{i\alpha} = -M\omega^2 x_1 + ic\omega x_1 + kx_1. \tag{4.87}$$

This equation illustrates that the total force phasor is the vector sum of contributions from the spring, damper, and mass inertia, each carrying a distinct phase shift.

The corresponding phasor diagram shows these components resolved along orthogonal axes: the spring force $F_s = kx_1$ lies along the real axis, while the damping $F_d = ic\omega x_1$ and inertial force $F_i = -M\omega^2 x_1$ are aligned with the imaginary axis. The total applied force phasor F_a therefore satisfies

$$F_a = F_s + F_d + F_i, \tag{4.88}$$

and its orientation relative to the real axis gives the phase lag α between force and displacement. This phase relationship is critical in determining the energy transfer efficiency, particularly in oscillatory cryocooler systems where force and displacement must be tightly synchronised. This decomposition is illustrated in figure 4.10, where each force component is represented as a vector in the complex plane, and their vector sum yields the applied force F_a. The diagram also highlights the phase lag α, which plays a critical role in quantifying how energy is transferred between the source and the oscillator.

Time-domain consistency: Expanding equation (4.87) into the time domain reconfirms the familiar form of the driven response:

$$x(t) = x_1\cos(\omega t), \tag{4.89}$$

$$\dot{x}(t) = -x_1\omega\sin(\omega t), \tag{4.90}$$

$$\ddot{x}(t) = -x_1\omega^2\cos(\omega t). \tag{4.91}$$

Substituting these into the original equation of motion yields

$$F_1\cos(\omega t + \alpha) = -M\omega^2 x_1\cos(\omega t) + c\omega x_1\sin(\omega t) + kx_1\cos(\omega t), \tag{4.92}$$

which confirms that the phasor formulation correctly captures the steady-state behaviour.

The utility of this representation becomes particularly apparent in section 4.3, where it underpins impedance-based modelling of cryocooler subsystems. It also generalises naturally to distributed systems (such as pulse tubes and regenerators), enabling concise expressions for pressure–volume work and phase-dependent flow distributions.

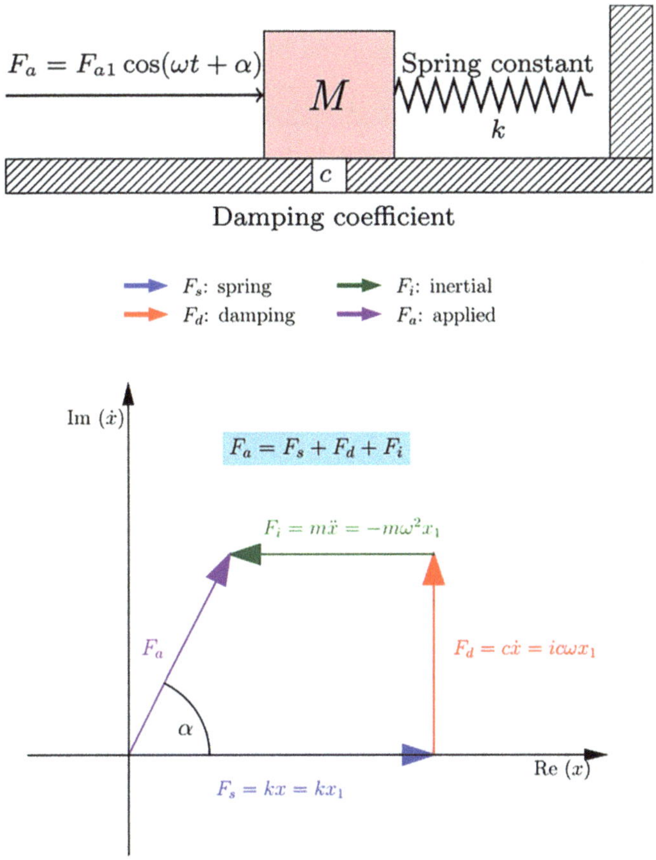

Figure 4.10. Combined schematic and phasor diagram of a damped, driven harmonic oscillator. The upper diagram illustrates the physical system comprising a mass M, spring constant k, and damping coefficient c, subjected to an external forcing term $F_a = F_{a1}\cos(\omega t + \alpha)$. The lower diagram presents the corresponding phasor representation in the complex plane, where the total applied force F_a is decomposed into contributions from the spring force F_s, damping force F_d, and inertial force F_i. The phase angle α denotes the lag between the applied force and the resulting displacement response. This visualisation highlights the vectorial nature of dynamic force balance in the frequency domain.

4.3.6.8 Harmonic analysis of a cryocooler compressor

Building on the phasor formulation for damped oscillatory systems, we now extend the analysis to the cryocooler compressor. Here, the moving piston behaves as a harmonically driven mass–spring–damper system, but with additional complexity introduced by dynamic pressure forces arising from gas compression. The system is driven by a sinusoidal electromagnetic force, and its response is governed by the superposition of elastic, damping, inertial, and pressure-related forces. Harmonic analysis using complex phasors provides a compact and insightful way to describe this behaviour.

Key design parameters required for impedance matching include the stroke amplitude x_1, spring constant k, damping coefficient c, piston area A, and moving mass M. These parameters collectively determine the dynamic response of the compressor. The natural frequency is given by

$$f_0 = \frac{1}{2\pi}\sqrt{\frac{k}{M}}. \tag{4.93}$$

In addition to the mechanical properties, motor parameters such as the force constant α, coil resistance R, and maximum allowable current $I_{\max}$ play a critical role in defining the electrical power requirements and thermal constraints of the compressor (Radebaugh 2015).

The compressor's thermodynamic efficiency η_{co} can be defined as the ratio of PV work to total electrical input:

$$\eta_{\mathrm{co}} = \frac{\dot{W}_{PV}}{\dot{W}_{el}} = \frac{\dot{W}_m - \dot{W}_d}{\dot{W}_m + \frac{1}{2}|I|^2 R} \tag{4.94}$$

where

$$\dot{W}_{\mathrm{m}} = \frac{1}{2}\alpha\omega|x|\cos\phi_{\mathrm{m}}$$
$$\dot{W}_{\mathrm{d}} = \frac{1}{2}c\omega^2 x^2.$$

At resonance ($\phi_{\mathrm{m}} = 0$), mechanical power transfer is maximised. Introducing the Joule heating frequency

$$f_{\mathrm{J}} = \frac{|I|R}{\pi\alpha s}, \quad s = 2|x| \tag{4.95}$$

we express compressor efficiency as

$$\eta_{\mathrm{co}} = \frac{1}{1 + (f_{\mathrm{J}}/f)} \tag{4.96}$$

indicating that minimising f_{J} and operating at high frequencies enhances efficiency. A typical operating range is 10 Hz $< f_{\mathrm{J}} <$ 30 Hz (Radebaugh 2015).

4.4 Summary

Phasor analysis provides a powerful framework for modelling oscillatory dynamics in cryocooler systems, transforming differential equations into algebraic relations in the complex plane. By representing sinusoidal quantities as rotating vectors, we gain intuitive and quantitative insight into the interplay between applied, inertial, damping, and restoring forces. This methodology extends naturally from simple harmonic oscillators to the forced, damped motion of cryocooler pistons, enabling precise characterisation of resonance conditions, impedance matching, and compressor efficiency. The use of phasors is thus foundational in both the theoretical understanding and practical design of high-performance cryogenic systems.

References

Abolghasemi M A, Rana H, Stone R, Dadd M, Bailey P and Liang K 2020 Coaxial stirling pulse tube cryocooler with active displacer *Cryogenics* **111** 103143

Almtireen N, Brandner J J and Korvink J G 2020 Pulse tube cryocooler: phasor analysis and one-dimensional numerical simulation *J. Low Temp. Phys.* **199** 1179–97

Beranek L L 1993 *Acoustics* (College Park, MA: American Institute of Physics)

de Waele A T M 2019 Pulse tube refrigerator with a warm displacer: theoretical treatment in the harmonic approximation *Cryogenics* **100** 53–61

Kinsler L E, Frey A R, Coppens A B and Sanders J V 2000 *Fundamentals of Acoustics* 4th edn (New York: Wiley)

Kittel P, Kashani A, Lee J and Roach P 1996 General pulse tube theory *Cryogenics* **36** 849–57

Radebaugh R 2003 Thermodynamics of regenerative refrigerators *Generation of Low Temperature and Its Applications* vol 3 (Kamakura: Shonan Technology Center)

Radebaugh R 2015 Analysis of regenerative cryocoolers *Zhejiang University Workshop*

Radebaugh R and Shirron P 2018 Foundations of Cryocoolers *International Cryocoolers Conference* (Cryogenic Society of America)

Radebaugh R 2019 Cryocooler fundamentals *Cryogenic Engineering Conference* (Cryogenic Society of America)

Rana H, Abolghasemi M A, Stone R, Dadd M, Bailey P and Liang K 2020 Numerical modelling of a coaxial Stirling pulse tube cryocooler with an active displacer for space applications *Cryogenics* **106** 103048

Rana H, Abolghasemi M A, Stone R, Dadd M and Bailey P B 2021 A passive displacer for a Stirling pulse tube cryocooler *International Cryocooler Conference (Boulder, CO)*

Rana H, Bailey P B, Dadd M and Stone R 2022a Compressor stroke and frequency response using strain gauges *International Cryocooler Conference (Boulder, CO)*

Rana H, Bailey P B, Dadd M and Stone R 2022b Experimental fatigue analysis of cryocooler flexure springs *IOP Conf. Ser.: Mater. Sci. Eng. ICEC-ICMC Proceedings*

Vanapalli S, Lewis M, Gan Z and Radebaugh R 2007 120 Hz pulse tube cryocooler for fast cooldown to 50 K *Appl. Phys. Lett.* **90** 072504

Chapter 5

Numerical modelling

Cryocooler performance prediction, component design, and optimisation often rely on detailed numerical models to capture real-world nonlinearities, coupled phenomena, and practical constraints that lie beyond the scope of purely analytical techniques. In this chapter, we explore the numerical foundations of cryocooler simulation, beginning with general modelling strategies, progressing through established 1D methods such as those implemented in Sage software, and concluding with an overview of alternative and emerging computational approaches.

5.1 Approaches to cryocooler numerical modelling

Numerical modelling of cryocoolers requires a careful formulation of the conservation laws that govern oscillatory fluid dynamics, thermodynamic energy exchange, and the mechanical behaviour of various interacting components. The systems are inherently nonlinear and involve complex coupling between pressure oscillations, mass flow, heat transport, and occasionally structural deformations. The governing physics are typically encoded in partial differential equations (PDEs) and ordinary differential equations (ODEs), capturing conservation of mass, momentum, and energy, along with relevant boundary conditions and constitutive relations.

5.1.1 Governing equations and key assumptions

The compressible Navier–Stokes equations form the foundation of detailed flow simulations in cryocoolers. In one dimension, the continuity, momentum, and energy equations take the following general forms for a control volume:

$$\frac{\partial \rho}{\partial t} + \frac{\partial (\rho u)}{\partial x} = 0, \tag{5.1}$$

$$\frac{\partial (\rho u)}{\partial t} + \frac{\partial}{\partial x}(\rho u^2 + p) = \frac{\partial \tau}{\partial x}, \tag{5.2}$$

doi:10.1088/978-0-7503-4826-3ch5

$$\frac{\partial E}{\partial t} + \frac{\partial}{\partial x}[(E + p)u] = \frac{\partial}{\partial x}(k\frac{\partial T}{\partial x}) + \Phi, \tag{5.3}$$

where ρ is the density, u the velocity, p the pressure, τ the viscous shear stress, E the total energy per unit volume, k the thermal conductivity, T the temperature, and Φ the viscous dissipation function. These equations are typically closed using an appropriate equation of state such as the ideal gas law

$$p = \rho RT, \tag{5.4}$$

where R is the specific gas constant. For helium and other working fluids used in cryocoolers, temperature-dependent properties must be considered (Nellis and Klein 2021).

5.1.2 Time-domain versus frequency-domain modelling

Cryocooler systems exhibit inherently oscillatory behaviour, particularly in regenerative and pulse tube coolers where gas flow and pressure oscillate harmonically at driving frequencies typically in the range of 20–100 Hz. Time-domain modelling solves the governing equations (5.1)–(5.3) directly in the temporal dimension, allowing the resolution of transient start-up, nonlinear wave interactions, and large-signal behaviour. Numerical methods such as the finite-difference method (FDM), finite volume method (FVM), or method of lines are often employed, with explicit or implicit time-stepping schemes.

A simple time-domain model for the pressure drop Δp in a laminar oscillating flow through a porous regenerator, based on Darcy–Forchheimer drag, can be written as

$$\Delta p(t) = \mu\frac{L}{K}u_s(t) + \beta\rho\frac{L}{K^{1/2}}u_s(t)^2, \tag{5.5}$$

where μ is the dynamic viscosity, L the regenerator length, K the permeability, β a flow resistance coefficient, and $u_s(t)$ is the superficial velocity.

Frequency-domain modelling assumes steady periodicity and represents variables as complex phasors. This transforms time derivatives into algebraic multipliers:

$$\frac{\mathrm{d}}{\mathrm{d}t} \to j\omega, \tag{5.6}$$

where ω is the angular frequency. Applying this to equation (5.1) and linearising for small perturbations yields

$$j\omega\hat{\rho} + \frac{\mathrm{d}}{\mathrm{d}x}\left(\rho_0\hat{u}\right) = 0, \tag{5.7}$$

where hatted quantities are complex phasor amplitudes, and ρ_0 is the mean density. Frequency-domain models significantly reduce computational cost and are ideally suited to performance prediction in the steady-state regime (Radebaugh 2009).

5.1.3 Lumped versus distributed parameter modelling

The modelling approach may also be categorised according to the spatial resolution of the variables:

5.1.3.1 Lumped parameter models

These assume uniformity of thermodynamic variables within each component and reduce PDEs to systems of ODEs. For example, the thermal mass of a component may be represented by a single-node energy balance:

$$C\frac{\mathrm{d}T}{\mathrm{d}t} = \dot{Q}_{\mathrm{in}} - \dot{Q}_{\mathrm{out}}, \tag{5.8}$$

where C is the heat capacity, and $\dot{Q}_{\mathrm{in}}$, $\dot{Q}_{\mathrm{out}}$ are incoming and outgoing heat rates. Lumped models enable rapid system-level simulations and are often employed in early design phases (Fereday *et al* 2006).

5.1.3.2 Distributed parameter models

These models account for spatial variation along the length of components. In regenerators, for instance, the enthalpy balance for the working gas and matrix are described by coupled PDEs:

$$\rho_{\mathrm{g}} c_{\mathrm{pg}} \frac{\partial T_{\mathrm{g}}}{\partial t} + \rho_{\mathrm{g}} u_{\mathrm{g}} c_{\mathrm{pg}} \frac{\partial T_{\mathrm{g}}}{\partial x} = hA_{\mathrm{s}}(T_{\mathrm{s}} - T_{\mathrm{g}}), \tag{5.9}$$

$$\rho_{\mathrm{s}} c_{\mathrm{ps}} \frac{\partial T_{\mathrm{s}}}{\partial t} = hA_{\mathrm{s}}(T_{\mathrm{g}} - T_{\mathrm{s}}), \tag{5.10}$$

where T_{g} and T_{s} are the gas and solid temperatures, c_{pg} and c_{ps} are specific heats, h is the heat transfer coefficient, and A_{s} is the specific surface area. These equations capture axial heat transfer and regenerator ineffectiveness, which are critical for accurate performance estimation (Nellis and Klein 2021).

Distributed models may also incorporate spatially resolved pressure losses using Darcy–Weisbach friction:

$$\frac{\mathrm{d}p}{\mathrm{d}x} = -f\frac{\rho u^2}{2D}, \tag{5.11}$$

where f is the friction factor and D is the hydraulic diameter. This is especially important in oscillating flows where minor losses accumulate over each cycle.

5.1.4 Thermal loss mechanisms and model extensions

Advanced models incorporate thermal parasitics such as shuttle heat transfer in regenerators, conduction through structural supports, and imperfect insulation. Shuttle heat transfer, for instance, can be represented as

$$\dot{Q}_{\mathrm{sh}} = \omega A_{\mathrm{s}} \Delta x \left(\frac{\mathrm{d}T}{\mathrm{d}x}\right), \tag{5.12}$$

where Δx is the oscillation amplitude. Losses in the pulse tube due to streaming and secondary flow vortices may also be included using semi-empirical damping factors or loss terms fitted to experimental data (Gedeon 1995b, Radebaugh 2000).

Ultimately, the choice between modelling approaches depends on the purpose of the simulation, desired accuracy, and computational budget. Frequency-domain, distributed models, such as those implemented in Sage, are often optimal for steady-state performance prediction, while time-domain, full Navier–Stokes models are favoured in research settings where transient behaviour, start-up phenomena, or turbulence are of interest.

5.2 1D modelling and Sage software

One of the most widely used tools for modelling cryocoolers is Sage, a 1D simulation platform specifically developed for oscillating-flow cryogenic systems (Gedeon 1998b). It provides a frequency-domain representation of thermofluid dynamics, ideal for simulating regenerative cryocoolers, pulse tube refrigerators, and hybrid architectures. Unlike general-purpose CFD or multiphysics solvers, Sage is optimised to model the oscillatory, low-Mach, high-Prandtl number flow regimes typically encountered in cryocoolers.

The software is structured around an object-oriented framework, where each subsystem is modelled as a chain of interconnected components (e.g. pistons, regenerators, heat exchangers, tubes, reservoirs), with each component solving the conservation laws of mass, momentum, and energy along a single spatial axis. Components communicate with adjacent ones by exchanging complex phasor values for pressure, flow rate, enthalpy, and acoustic power.

5.2.1 Modelling philosophy

Sage assumes steady-periodic operation at a single frequency, which simplifies the governing partial differential equations (PDEs) into a set of complex-valued algebraic equations. These are solved using iterative frequency-domain techniques. Time-varying quantities such as pressure $p(t)$ and mass flow rate $\dot{m}(t)$ are represented as complex phasors:

$$p(t) = \Re\{\hat{p}e^{i\omega t}\}, \tag{5.13}$$

$$\dot{m}(t) = \Re\{\hat{\dot{m}}e^{i\omega t}\}, \tag{5.14}$$

where $\hat{p}$ and $\hat{\dot{m}}$ are the complex pressure and mass flow amplitudes, respectively, and ω is the angular frequency. This transformation reduces the temporal derivatives to multipliers:

$$\frac{\mathrm{d}}{\mathrm{d}t} \rightarrow i\omega, \tag{5.15}$$

allowing the continuity and momentum equations to be written in steady-harmonic form.

5.2.1.1 Mass conservation

In the 1D phasor domain, the continuity equation becomes

$$i\omega\hat{\rho} + \frac{\mathrm{d}}{\mathrm{d}x}\left(\rho_0\hat{u}\right) = 0, \tag{5.16}$$

assuming small perturbations around a mean flow ρ_0. This formulation ensures that all mass inflow and outflow are phase-consistent across elements.

5.2.1.2 Momentum conservation

The phasor form of the linearised momentum equation is given by

$$i\omega\rho_0\hat{u} + \frac{\mathrm{d}\hat{p}}{\mathrm{d}x} + R_\mathrm{f}\hat{u} = 0, \tag{5.17}$$

where R_f is a complex flow resistance coefficient that includes both viscous and inertial effects. Empirical corrections are often applied to R_f to capture non-sinusoidal distortion, end effects, and minor losses (Gedeon 1998b, Radebaugh 2000).

5.2.1.3 Energy conservation

The energy equation is handled through enthalpy flows and thermal diffusion terms For the gas side in a regenerator, the governing equation becomes

$$i\omega\rho_\mathrm{g}c_\mathrm{pg}\hat{T}_\mathrm{g} + \rho_\mathrm{g}u_\mathrm{g}c_\mathrm{pg}\frac{\mathrm{d}\hat{T}_\mathrm{g}}{\mathrm{d}x} = h_\mathrm{s}a_\mathrm{s}\left(\hat{T}_\mathrm{s} - \hat{T}_\mathrm{g}\right), \tag{5.18}$$

where h_s is the gas–solid heat transfer coefficient, a_s is the specific surface area per unit volume, and $\hat{T}_\mathrm{g}$, $\hat{T}_\mathrm{s}$ are the complex temperature phasors for the gas and matrix, respectively. This equation is coupled to the matrix energy equation

$$i\omega\rho_\mathrm{s}c_\mathrm{ps}\hat{T}_\mathrm{s} = h_\mathrm{s}a_\mathrm{s}\left(\hat{T}_\mathrm{g} - \hat{T}_\mathrm{s}\right) + k_\mathrm{s}\frac{\mathrm{d}^2\hat{T}_\mathrm{s}}{\mathrm{d}x^2}, \tag{5.19}$$

where k_s is the thermal conductivity of the regenerator matrix. These equations form a coupled system that captures the regenerator effectiveness and phase lag between pressure and enthalpy flow.

5.2.1.4 Acoustic power

The acoustic power transferred through a cross-section is defined by

$$\dot{W}_\mathrm{ac} = \frac{1}{2}\Re\{\hat{p}\,\hat{U}^*\}, \tag{5.20}$$

where $\hat{U}$ is the complex volumetric flow rate and the asterisk denotes the complex conjugate. Acoustic power is a key performance metric in Sage, in particular for evaluating losses across components and identifying sources of parasitic dissipation.

5.2.1.5 Impedance relations

Many components in Sage are modelled as lumped impedance elements, with pressure–flow relationships of the form

$$\hat{p} = Z\hat{U}, \tag{5.21}$$

where Z is a complex impedance that can include resistive, inertial, and capacitive (compliance) contributions. For example, a straight tube of length L and cross-sectional area A filled with a gas of dynamic viscosity μ has an acoustic impedance:

$$Z = \frac{i\omega\rho_0 L}{A} + \frac{8\mu L}{\pi r^4}, \tag{5.22}$$

where r is the radius of the tube. These relations enable fast, modular simulation of multi-component systems.

5.2.2 Thermodynamic loss models

One of the critical strengths of Sage as a modelling tool lies in its incorporation of empirical and semi-empirical models that account for thermodynamic irreversibilities and parasitic losses intrinsic to real cryocooler systems. These losses, while often secondary to ideal thermodynamic performance in conceptual models, play a dominant role in limiting the actual efficiency and cooling power of deployed devices. As such, the inclusion of detailed loss mechanisms allows Sage to more accurately predict net performance and to serve as a design optimisation platform.

A key source of inefficiency in regenerative cryocoolers is regenerator ineffectiveness, which arises due to imperfect enthalpy exchange between the oscillating working gas and the solid matrix. In a well-functioning regenerator, the gas and matrix undergo nearly out-of-phase thermal excursions, allowing the gas to recover its thermal energy with minimal entropy generation. However, in practice, finite heat transfer coefficients, axial conduction, and limited thermal penetration depth reduce the effectiveness of this exchange. The resulting net loss manifests as an elevated temperature span across the regenerator and a corresponding drop in cooling power at the cold end. The regenerator effectiveness ε is often modelled as a function of dimensionless groups such as the Stanton number (St), Reynolds number (Re), and matrix Prandtl number (Pr_s), with performance degradation linked to the ratio of oscillation period to thermal penetration depth (Gedeon 1995a).

Another dominant parasitic mechanism captured by Sage is shuttle heat transfer, which occurs when an oscillating component (such as a displacer or piston) moves between hot and cold regions without direct fluid exchange. This mechanism introduces a back and forth conduction pathway that bypasses the working gas, acting as a thermal short. In Sage, the shuttle heat transfer rate is approximated using the expression

$$\dot{Q}_{\mathrm{sh}} = k\,\omega x_0 A_{\mathrm{s}} \frac{\mathrm{d}T}{\mathrm{d}x}, \tag{5.23}$$

where ω is the angular frequency of oscillation, x_0 is the peak stroke amplitude of the shuttle component, A_s is the effective cross-sectional area participating in heat transfer, and dT/dx is the local axial temperature gradient. This formulation assumes harmonic oscillation and neglects transient conduction lags, providing a tractable, first-order estimate of the shuttling contribution. Shuttle losses become particularly severe in geometries with narrow gaps and large temperature gradients, such as regenerators coupled to actively driven displacers (Radebaugh 2000).

Flow disturbances caused by abrupt changes in geometry, such as elbows, T-junctions, sudden contractions, or expansions, also lead to energy dissipation in the form of minor losses. These are incorporated into Sage via the introduction of effective loss coefficients K_e that augment the pressure drop in the momentum equation. Specifically, the phasor form of the momentum equation (see equation (5.17)) is modified to include an additional term of the form

$$\Delta p_{\text{minor}} = \frac{1}{2} K_e \rho_0 |\hat{u}|^2, \tag{5.24}$$

where K_e is empirically determined based on component geometry and flow regime, and $|\hat{u}|$ is the amplitude of the oscillatory velocity. These coefficients are often drawn from standard handbooks on fluid mechanics but can be calibrated using experimental data. Since oscillatory flow lacks a well-defined Reynolds number, corrections are applied to account for phase-lagged inertia and energy storage in the boundary layer (Gedeon 1998b).

A further important loss mechanism arises from streaming flows in pulse tube cryocoolers. Unlike purely harmonic acoustic flows, streaming is a time-averaged mass transport driven by nonlinearity in the velocity field, thermoviscous effects, and asymmetries in geometry or thermal boundary conditions. This steady flow contributes to entropy generation and can carry heat from the hot end of the pulse tube back toward the cold end, severely degrading performance. Sage incorporates simplified streaming models based on flow asymmetry and secondary flow maps, with the streaming power loss typically modelled as an effective heat load:

$$\dot{Q}_{\text{stream}} = \rho_0 \bar{u} c_p (T_H - T_C), \tag{5.25}$$

where $\bar{u}$ is the mean (nonzero) streaming velocity, c_p is the specific heat at constant pressure, and T_H and T_C are the hot and cold-end temperatures, respectively. The streaming term is often estimated from the second-order expansion of the Navier–Stokes equations or from empirical correlations developed from experimental data (Gedeon 1998b).

Additional thermodynamic losses modelled in Sage include axial conduction through the solid matrix and envelope walls, regenerator bypass leakage, imperfect heat exchanger coupling, and gas property variation at cryogenic temperatures. These are generally included through embedded material models and semi-empirical subroutines that adjust the effective thermal conductance, heat transfer coefficients, or flow resistance as a function of local state variables.

Taken together, these thermodynamic loss models are crucial for translating theoretical cycle performance into realistic engineering expectations. While they may introduce increased modelling complexity, their integration enables Sage to serve not just as a predictive tool but as a means of guiding component-level design trade-offs and full-system optimisation. As demonstrated by numerous experimental validations (Choi *et al* 2004, Nellis and Klein 2021), the fidelity of these loss models is sufficient to capture most real-world inefficiencies within a few percent of experimental measurements, making Sage a valuable tool in both research and industrial development.

5.2.3 Model assembly and convergence

The process of assembling a model in Sage begins with constructing a nodal diagram that represents the physical layout of the cryocooler in terms of interconnected functional components. This schematic captures the topological and thermodynamic relationships between elements such as compressors, regenerators, pulse tubes, heat exchangers, inertance tubes, and buffer volumes. Each node in the diagram corresponds to a control volume or flow junction, while each link represents a specific component model governed by its own set of constitutive and conservation equations.

Model construction in Sage follows a modular paradigm, enabling users to build up increasingly complex systems from basic building blocks. For instance, a compressor is linked via a tube to a warm heat exchanger, which then connects to a regenerator, followed by a cold heat exchanger, a pulse tube, and ultimately an inertance tube and reservoir. Each element introduces its own impedance and phase relationships into the system, and the nodal diagram governs how these relationships are coupled. Boundary nodes are used to impose known pressure amplitudes, thermal loads, or drive frequencies, while internal nodes enforce mass and energy continuity.

Once the model is assembled, Sage transforms the governing equations into a global set of nonlinear algebraic equations expressed in the frequency domain. Each unknown phasor variable (e.g. pressure amplitude $\hat{p}$, volumetric flow $\hat{U}$, enthalpy flow $\hat{H}$) at each node or component interface becomes part of a large complex-valued system:

$$\mathbf{F}(\mathbf{X}) = 0, \tag{5.26}$$

where $\mathbf{X}$ is the vector of all unknown phasors and $\mathbf{F}$ represents the vector of residuals resulting from applying conservation laws and constitutive relations across all components. The size of $\mathbf{X}$ scales with the number of interfaces and phasor variables tracked per node.

To solve this nonlinear system, Sage employs iterative root-finding methods such as the Newton–Raphson or quasi-Newton Broyden algorithm. In the Newton–Raphson method, convergence is achieved by linearising the residual equations about a current guess $\mathbf{X}_k$ using a Jacobian matrix $\mathbf{J}$:

$$\mathbf{X}_{k+1} = \mathbf{X}_k - \mathbf{J}^{-1}\mathbf{F}(\mathbf{X}_k). \tag{5.27}$$

Broyden's method avoids repeated computation of the Jacobian by updating it approximatively from successive iterations. These solvers are robust for moderately sized systems and converge efficiently when good initial guesses are provided. Convergence is typically monitored based on residual norms (e.g. $\|\mathbf{F}\|_{\infty} < 10^{-6}$) or the maximum relative change in any phasor variable across iterations.

Sage offers extensive built-in diagnostics for interpreting simulation outcomes, enabling detailed insight into both global performance and localised thermodynamic behaviour. Among the most informative outputs are phasor plots, which show the pressure and volumetric flow amplitude as a function of position along the cryocooler, providing a clear visual representation of wave propagation and impedance mismatches. The software also tracks acoustic power and enthalpy flux throughout the system, allowing engineers to identify where energy is transmitted, dissipated, or reflected. In addition, Sage quantifies loss mechanisms by computing the entropy generation in each component, offering a thermodynamically rigorous basis for evaluating inefficiencies. Finally, it reports key time-averaged performance metrics, including the net cooling power at the cold heat exchanger, the total electrical input power, and the overall thermodynamic efficiency of the cycle, enabling comprehensive performance assessment and optimisation.

These diagnostic capabilities allow users to trace inefficiencies, identify mismatches in impedance, and adjust component dimensions or operating conditions to improve performance. The transparency of the nodal architecture also facilitates parametric sweeps, sensitivity analyses, and rapid prototyping of alternative system topologies.

5.2.4 Phasor node network and governing equations

In Sage's frequency-domain framework, the cryocooler is modelled as a network of components connected via discrete nodes. Each component, such as a regenerator, pulse tube, or inertance tube, has internal equations that govern its phasor behaviour, typically relating input and output pressure, flow rate, and enthalpy through impedance relations and energy balances. However, the overall physical consistency of the system is not enforced at the component level alone. Instead, it is the role of the *nodes* to ensure global thermodynamic conservation across the network.

A node in this context is a junction point at which two or more components interface. At each node, Sage enforces complex-valued conservation laws: namely, mass conservation via volumetric flow, energy conservation via enthalpy flow, and acoustic power balance. These are implemented as algebraic residual equations that must sum to zero at each node, ensuring that there is no unphysical creation or destruction of mass, energy, or power across the system.

The phasor representation simplifies the time-dependent conservation equations into frequency-domain constraints. The conservation of mass is expressed by requiring the complex volumetric flow rates from all N components connected to the node to sum to zero:

$$\sum_{k=1}^{N} \hat{U}_k = 0, \tag{5.28}$$

where $\hat{U}_k$ is the phasor volumetric flow entering or exiting the node from component k. This ensures continuity of oscillating flow in the frequency domain, analogous to Kirchhoff's current law in AC circuits.

Similarly, Sage tracks the flow of oscillatory mechanical energy, in other words, the acoustic power, by enforcing that the sum of real acoustic power from all connected branches also vanishes:

$$\sum_{k=1}^{N} \Re\left\{\hat{p}_k \hat{U}_k^*\right\} = 0, \tag{5.29}$$

where $\hat{p}_k$ is the pressure phasor at interface k, $\hat{U}_k^*$ is the complex conjugate of the flow phasor, and the product gives the local acoustic power contribution. This equation ensures that the net input and output power at the node are balanced, barring explicit loss terms accounted for inside components.

Finally, the conservation of thermal energy across the node is ensured by requiring the sum of phasor enthalpy flows to be zero:

$$\sum_{k=1}^{N} \hat{H}_k = 0, \tag{5.30}$$

where $\hat{H}_k = \hat{m}_k \hat{h}_k$ represents the phasor enthalpy flow from each component, calculated from the complex mass flow rate $\hat{m}_k$ and specific enthalpy $\hat{h}_k$.

It is important to emphasise that these conservation equations are enforced only at the nodes, not within the components themselves. Each component is responsible solely for computing its own output phasor quantities (e.g. $\hat{p}_k$, $\hat{U}_k$, $\hat{H}_k$) based on its internal thermofluidic and structural model. These values are then passed to the adjacent nodes, which assemble them into global residuals to check whether the conservation constraints are satisfied.

From the solver's perspective, the entire cryocooler model is thus built from two classes of equations: internal component relations and nodal conservation laws. Together, these form a coupled nonlinear system in which each node contributes three residuals, one each for mass, acoustic power, and enthalpy. The Sage solver iteratively adjusts all unknown phasor quantities to drive the residuals at every node toward zero. This structure allows components to be modelled flexibly and independently, while the global network retains strict thermodynamic and acoustic consistency.

Figure 5.1 illustrates a typical phasor node configuration used in one-dimensional frequency-domain cryocooler simulations. In this abstraction, each component (labelled A through D) connects to a central node and contributes its local phasor quantities: complex volumetric flow rate $\hat{U}_k$, pressure $\hat{p}_k$, and enthalpy flow $\hat{H}_k$. These are computed internally by each component model and passed to the node for global balancing. The node itself does not model dynamics, but rather enforces three core conservation principles: mass conservation, enthalpy conservation, and real acoustic power continuity. These appear as phasor summation constraints applied across all connected components. As part of the numerical solution process, these node residuals are driven toward zero by an iterative nonlinear solver. The diagram

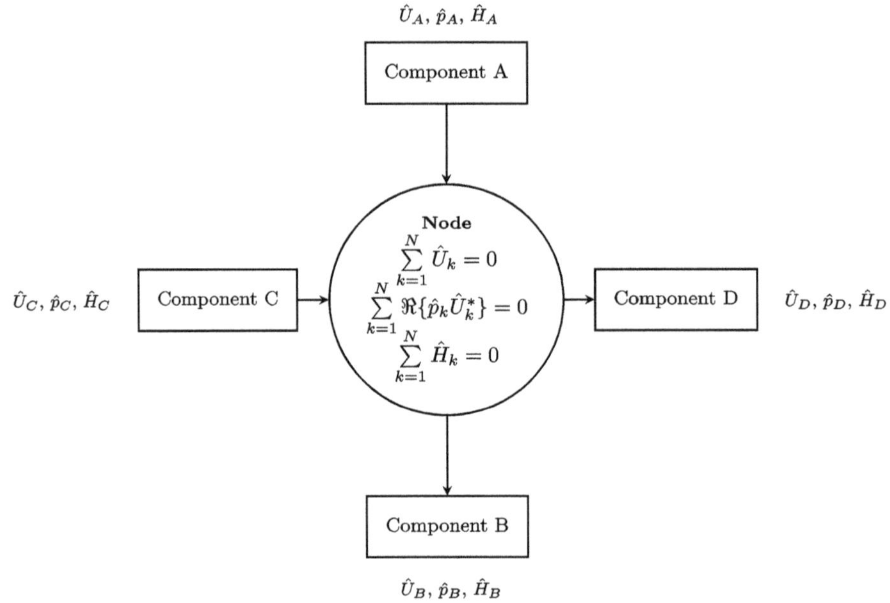

Figure 5.1. Phasor-node diagram illustrating the coupling between four cryocooler components (A–D) through a shared node. Each component provides complex phasor quantities for volumetric flow rate $\hat{U}_k$, pressure $\hat{p}_k$, and enthalpy flow $\hat{H}_k$. The central node enforces three conservation laws: mass continuity, acoustic power balance, and enthalpy conservation. Arrows indicate the direction of modelled signal flow and not necessarily net physical flow.

is representative of how full cryocooler networks are constructed in software such as Sage, where dozens of such nodes may be distributed across complex geometries.

5.2.5 Simulation workflow

The typical simulation workflow in Sage begins with the construction of a detailed schematic representing the cryocooler architecture. Each subsystem, such as the compressor, regenerator, pulse tube, heat exchangers, inertance tube, and reservoir, is modelled using a series of one-dimensional components, interconnected through phasor-based flow and pressure nodes. The user begins by defining a component-level layout using Sage's visual interface or scripting environment, ensuring that all components are properly connected through compatible boundary conditions.

Once the physical architecture is defined, the user specifies key system-wide parameters. These include the thermophysical properties of the working fluid (for example helium at cryogenic pressures), the oscillation frequency f (typically 30–60 Hz for Stirling-type machines), and thermal boundary conditions at the warm and cold ends.

In a Sage simulation, several numerical parameters must be defined prior to execution to ensure stable and accurate solutions. These include the axial mesh resolution Δx within distributed components such as regenerators and pulse tubes, which determines the spatial granularity of the finite-difference discretisation.

Another critical parameter is the convergence tolerance ε used by iterative solvers such as Newton–Raphson or Broyden methods; this is typically set between 10^{-6} and 10^{-8} to balance numerical stability with computational cost. Additionally, the user must specify the maximum harmonic number n to be included in the Fourier expansion of phasor quantities. For instance, pressure and volumetric flow rate are expressed as

$$\hat{p}(x, t) = \sum_{n=1}^{N} P_n(x)\mathrm{e}^{\mathrm{i}n\omega t}, \quad \hat{U}(x, t) = \sum_{n=1}^{N} U_n(x)\mathrm{e}^{\mathrm{i}n\omega t}, \tag{5.31}$$

where $\omega = 2\pi f$ is the angular frequency of the system. The choice of N determines the spectral resolution of the simulation and directly impacts the fidelity with which nonlinear interactions and waveform distortions are captured.

The solution process then proceeds by assembling the full system of governing equations across all nodes and components. Each node enforces conservation of mass, enthalpy, and acoustic power (see equations (5.16)–(5.30)), while each component applies momentum balance, thermal conduction, and empirical loss terms For example, in a generic duct, the frequency-domain momentum equation is

$$\mathrm{i}\omega\rho_0 U + \frac{\partial \hat{p}}{\partial x} + f_\mathrm{D}\frac{\rho_0 U|U|}{2D_\mathrm{h}} = 0, \tag{5.32}$$

where ρ_0 is the mean density, D_h is the hydraulic diameter, and f_D is a Darcy–Weisbach friction factor.

Similarly, in components with heat transfer (e.g. regenerators or exchangers), the axial energy equation is solved in steady-periodic form:

$$\mathrm{i}\omega T + u\frac{\mathrm{d}T}{\mathrm{d}x} = \alpha\frac{\mathrm{d}^2 T}{\mathrm{d}x^2} - \frac{q'''}{\rho c_\mathrm{p}}, \tag{5.33}$$

where α is the thermal diffusivity and q''' represents volumetric heating or cooling.

After defining all inputs, the user runs the solver. Sage performs phasor-based calculations and iteratively converges the solution. Sage solves at the fundamental ($N = 1$); multi-harmonic treatments require extended/harmonic-balance models. The residual at each node is monitored using:

$$R_i^{(k)} = \left|\sum_{j=1}^{N} \hat{U}_{ij}^{(k)}\right|, \tag{5.34}$$

where k denotes the iteration number. Convergence is achieved when all residuals fall below a predefined threshold ε. Figure 5.2 illustrates a complete Sage model of a single-stage pulse tube cryocooler. The model is broken into labelled subsystems representing physical groupings in the cooler: the displacer unit (**DISP**), connecting tubes (**TUB**), regenerator and warm-end heater block (**REG**, **CLR**), pulse tube (**PT**), warm heat exchanger (**WHX**), and cold head region (**CHX**). Each box contains Sage components, such as flow straighteners, heat exchangers, and pressure-controlled pistons. The arrows denote phasor-based mass flow (`mGtf`) and enthalpy flow

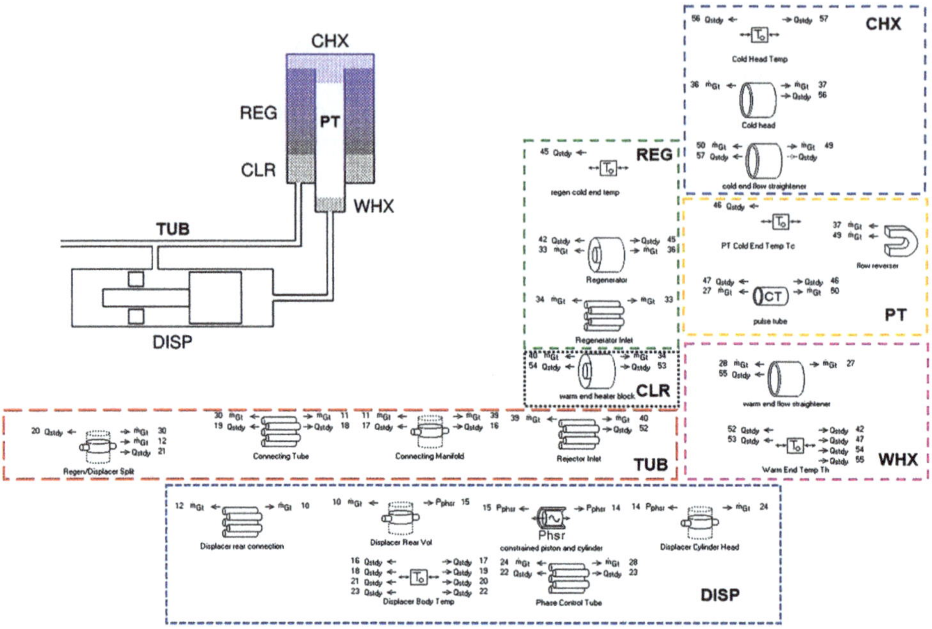

Figure 5.2. A Sage model schematic of a portion of a Stirling-pulse tube cryocooler (here, the warm-end displacer and cold-end assembly is featured). Each labelled block (e.g. DISP, TUB, PT, WHX) represents a subsystem comprising one or more components. Arrows indicate mass and energy flow between sections, and internal elements represent individual Sage components such as regenerators, flow straighteners, and pulse tubes.

(`Qtdy`) between components, allowing visual tracking of thermal and acoustic energy across the system. By organising the model this way, Sage enables modular analysis, visualisation, and parametric tuning of each subassembly while maintaining an interconnected system-wide solution.

Sage also provides rich post-processing capabilities. Diagnostic tools include phasor plots of pressure and velocity, entropy generation maps, component efficiency trends, and acoustic power flow visualisations. Users can perform parametric sweeps over frequency, charge pressure, or geometry to explore sensitivity and guide optimisation. For large design studies, Sage's scripting interface allows automated batch simulations, which is especially useful in multi-objective design (Radebaugh 2009, Nellis 2021).

5.2.6 Distributed component physics and empirical correlations

While previous sections described the high-level network formulation and phasor-based conservation at cryocooler nodes, Sage's physical realism hinges on accurate treatment of distributed elements such as regenerators, pulse tubes, and woven screens. These components span significant axial lengths and exhibit spatially varying thermofluid behaviour, necessitating localised differential modelling. This section expands on the internal physics and empirical correlations used to model such components, focusing on pressure loss, flow morphology, effective conductivity, and heat transfer approximations.

A principal performance determinant in regenerative cryocoolers is the woven mesh matrix, commonly fabricated from stainless steel or phosphor-bronze wire. Its geometry, porosity β, wire diameter d, and block diameter D, critically influences permeability and pressure drop. Since real mesh blocks deviate from ideal packing due to compaction and thermal cycling, porosity often requires empirical calibration. The Darcy–Weisbach friction factor in such porous media is approximated using the Ergun correlation:

$$\Delta p = \frac{150\mu L}{d^2}\frac{(1-\beta)^2}{\beta^3}v_s + \frac{1.75L\rho}{d}\frac{1-\beta}{\beta^3}v_s|v_s|, \tag{5.35}$$

where v_s is the superficial velocity, μ the dynamic viscosity, ρ the gas density, and L the matrix length (Ergun 1952). This can be recast into a friction factor using an empirical form suited to oscillatory flow:

$$f_f = \frac{129}{\text{Re}} + 2.91 \cdot \text{Re}^{-0.103} \tag{5.36}$$

with the Reynolds number defined as

$$\text{Re} = \frac{\rho u D_\text{h}}{\mu}. \tag{5.37}$$

In Sage, such components are modelled using a staggered-grid finite-difference method. The discretised conservation equations for mass, momentum, and energy over a control volume are

$$\frac{\partial \rho A}{\partial t} + \frac{\partial \rho u A}{\partial x} = 0 \tag{5.38}$$

$$\frac{\partial \rho u A}{\partial t} + \frac{\partial u \rho u A}{\partial x} + \frac{\partial P}{\partial x}A - F_{\text{vp}}A = 0 \tag{5.39}$$

$$\frac{\partial \rho e A}{\partial t} + P\frac{\partial A}{\partial t} + \frac{\partial}{\partial x}(u\rho e A + uPA + q) - Q_\text{w} = 0. \tag{5.40}$$

The staggered formulation, where ρ and ρe are computed at cell centres and $\rho u A$ at boundaries, was developed by Gedeon (1995b) to maintain numerical conservation and stability during oscillatory operation. Viscous force and axial conduction terms are respectively modelled as

$$F_{\text{vp}} = -\left(\frac{f_\text{D}}{d_\text{h}} + \frac{K}{L}\right)\rho u\frac{|u|}{2} \tag{5.41}$$

$$q = -N_k k A\frac{\partial T}{\partial x}. \tag{5.42}$$

The dimensionless conductivity enhancement factor N_k adjusts for thermal losses induced by turbulence and axial conduction. For the pulse tube, where large thermal gradients exist, Gedeon introduced an extended compliance-duct model that partitions heat flux into constituent transport mechanisms (Gedeon 1995a):

$$\frac{q}{q_{\mathrm{m}}} = 1 + \frac{q_{\mathrm{t}}}{q_{\mathrm{m}}} + \frac{q_{\mathrm{f}}}{q_{\mathrm{m}}} + \frac{q_{\mathrm{b}}}{q_{\mathrm{m}}} + \frac{q_{\mathrm{s}}}{q_{\mathrm{m}}}. \tag{5.43}$$

Here, q_{m} is molecular conduction, q_{t} is turbulent, q_{f} is free convection, q_{b} is boundary convection, and q_{s} is streamline convection. Each mode becomes significant in different regions depending on local Rayleigh number and geometry.

Moreover, heat transfer through matrix materials is strongly affected by tortuosity. In Sage, Maxwell's effective medium theory, corrected with Gedeon's calibration factor, is used to calculate the effective tortuosity f_{s}:

$$f_{\mathrm{s}} = \left(\frac{k_{\mathrm{s}}}{k_{\mathrm{g}}}\right)^{m-1} \left[\frac{3\left(\frac{k_{\mathrm{s}}}{k_{\mathrm{g}}} - \beta\right) + \left(2 + \frac{k_{\mathrm{s}}}{k_{\mathrm{g}}}\right)\beta}{3(1 - \beta) + \left(2 + \frac{k_{\mathrm{s}}}{k_{\mathrm{g}}}\right)\beta}\right], \tag{5.44}$$

where k_{s} and k_{g} are solid and gas conductivities, and m is a material-dependent empirical exponent (Carslaw 1959, Muralidhar 2001). Gedeon (1998a) noted that accurate tortuosity models are essential to correctly represent acoustic impedance and regenerative thermal coupling.

Finally, in harmonically oscillating systems such as pulse tube refrigerators, the phasor formulation allows effective calculation of distributed impedance. For a regenerative matrix, the acoustic impedance per unit length Z_x is given by

$$Z_x = \frac{\partial \hat{p}}{\partial x} / \hat{U}, \tag{5.45}$$

where $\hat{p}$ and $\hat{U}$ are the pressure and volumetric flow rate phasors. Phase shift, pressure drop, and enthalpy flow are all resolved in the frequency domain, enabling highly efficient spectral solutions.

Taken together, these distributed models and empirical formulations provide the necessary fidelity for predicting dynamic losses, regenerating efficiency, and spatial performance characteristics within advanced cryocoolers. Their implementation in Sage, anchored in Gedeon's physical framework, has become a cornerstone of modern numerical cryogenic analysis.

5.2.7 Limitations

While powerful, Sage is inherently limited by its 1D assumption and idealised component libraries. It cannot model 2D/3D effects, turbulence, or large transients. It also depends on accurate empirical correlations, which may not generalise across novel designs or unconventional operating regimes.

5.3 Time-domain finite volume modelling

5.3.1 Finite volume methods

In addition to Sage's frequency-domain framework, time-domain finite volume (TD-FV) modelling offers a complementary numerical method to simulate the dynamic behaviour of Stirling cryocoolers. A notable implementation is presented by Rawlings (2022), who developed a 1D finite volume model for the transient simulation of oscillating gas flow through distributed components, including regenerators and pulse tubes. Unlike Sage, which assumes steady-state periodicity and harmonic behaviour, the TD-FV approach resolves the temporal evolution of conservation laws using an explicit or semi-implicit time integration scheme over a spatial mesh.

The TD-FV method solves the governing conservation equations for mass, momentum, and energy in conservative form:

$$\frac{\partial \rho}{\partial t} + \frac{\partial (\rho u)}{\partial x} = 0 \tag{5.46}$$

$$\frac{\partial (\rho u)}{\partial t} + \frac{\partial (\rho u^2 + P)}{\partial x} = -F_{\text{visc}} \tag{5.47}$$

$$\frac{\partial (\rho e)}{\partial t} + \frac{\partial [(P + \rho e)u]}{\partial x} = -Q_{\text{loss}}, \tag{5.48}$$

where ρ is the gas density, u is the axial velocity, P is pressure, e is the total energy per unit mass, F_{visc} is the viscous loss term, and Q_{loss} accounts for distributed heat losses including wall conduction and convective boundary terms. These equations are discretised using control volume averaging and updated iteratively in time.

Rawlings (2022) implemented this scheme using MATLAB, employing a collocated grid with staggered variable placement for improved numerical stability. The spatial domain was divided into uniform segments, and boundary conditions were enforced via characteristic-based methods to simulate inlet pressures or mass flows. A typical example of the discretised form for continuity is

$$\rho_i^{n+1} = \rho_i^n - \frac{\Delta t}{\Delta x}\left[(\rho u)_{i+1/2}^n - (\rho u)_{i-1/2}^n\right]. \tag{5.49}$$

This contrasts with Sage's harmonic formulation which expresses state variables as sums of complex phasors as shown in equation (5.31).

Sage thereby directly solves for the steady-periodic solution at a given frequency ω, avoiding time-stepping altogether (Gedeon 1998b).

While Sage solves the governing equations in the frequency domain assuming periodic steady state, an alternative and powerful approach is time-domain simulation via finite volume methods. These methods discretize the conservation equations of mass, momentum, and energy in both space and time, enabling direct simulation of transient behaviour, start-up response, and non-periodic flows.

One such implementation is the 1D time-domain MacCormack scheme developed by Rawlings (2022), which solves the full compressible Navier–Stokes equations using a predictor–corrector strategy.

Algorithm 1. MacCormack finite volume solver for 1D Stirling cryocooler flow.

Initialise domain: mesh points x_j, time step Δt, grid spacing Δx
Initialise fields: $\rho_j^0, u_j^0, p_j^0, T_j^0, E_j^0$ **for** *each time step* $n = 0, 1, 2, \ldots$ **do**
Predictor step (forward differencing): for $j = 1$ to $N - 1$ **do**
$\rho_j^* = \rho_j^n - \frac{\Delta t}{\Delta x}(\rho_j^n u_j^n - \rho_{j-1}^n u_{j-1}^n)$
$u_j^* = u_j^n - \frac{\Delta t}{\Delta x}(\frac{p_j^n - p_{j-1}^n}{\rho_j^n})$
$E_j^* = E_j^n - \frac{\Delta t}{\Delta x}(u_j^n(E_j^n + p_j^n) - u_{j-1}^n(E_{j-1}^n + p_{j-1}^n))$

Recompute primitive variables from predicted fields for $j = 1$ to $N - 1$
$T_j^* = f(E_j^*, u_j^*)$ ▷Based on equation of state
$p_j^* = \rho_j^* R T_j^*$

Corrector step (backward differencing): for $j = 1$ to $N - 1$
$\rho_j^{n+1} = \frac{1}{2}(\rho_j^n + \rho_j^* - \frac{\Delta t}{\Delta x}(\rho_{j+1}^* u_{j+1}^* - \rho_j^* u_j^*))$
$u_j^{n+1} = \frac{1}{2}(u_j^n + u_j^* - \frac{\Delta t}{\Delta x}(\frac{p_{j+1}^* - p_j^*}{\rho_j^*}))$
$E_j^{n+1} = \frac{1}{2}(E_j^n + E_j^* - \frac{\Delta t}{\Delta x}(u_{j+1}^*(E_{j+1}^* + p_{j+1}^*) - u_j^*(E_j^* + p_j^*)))$

Apply boundary conditions at $j = 0$ and $j = N$

This time-marching approach offers several advantages over frequency-domain tools such as Sage, particularly in capturing start-up transients, shock formation, or flow instabilities. However, the approach is computationally more demanding, requiring careful enforcement of stability conditions (e.g. CFL criterion), and fine resolution of fast time-scale acoustic effects. Rawlings (2022) validated the solver against experimental cold head temperature evolution and showed good agreement with Sage in steady-state limits.

5.3.1.1 General finite volume algorithm with CFL stability

The finite volume method (FVM) is a versatile approach for solving conservation laws in compressible flow systems, including those encountered in Stirling and pulse tube cryocoolers. A key consideration in time-dependent simulations using FVM is numerical stability, which is governed by the Courant–Friedrichs–Lewy (CFL) condition. The CFL criterion ensures that information does not propagate across more than one grid cell in a single time step, maintaining physical fidelity and numerical robustness.

The CFL condition can be expressed as

$$\mathrm{CFL} = \max_j \left(\frac{u_j \Delta t}{\Delta x} \right) < 1 \tag{5.50}$$

where u_j is the local fluid velocity, Δt is the simulation time step, and Δx is the spatial resolution of the grid. When this condition is violated, the solution may exhibit non-physical oscillations or numerical instability. For this reason, modern solvers often incorporate adaptive time-stepping to enforce this constraint dynamically.

The following algorithm provides a generalised framework for finite volume simulation of cryogenic flows, incorporating numerical flux computation, conservative updates, equation-of-state recovery, boundary enforcement, and CFL enforcement.

In practical implementations, CFL violations do not necessarily terminate the simulation. Instead, the solver adaptively reduces the time step until the stability condition is met, ensuring robust convergence across a wide range of transient conditions. This capability is particularly vital in modelling start-up flows, high-pressure transients, or regimes with steep gradients, which are common in space-based cryocooler systems.

Algorithm 2. Finite volume time-stepping with CFL enforcement.

Input: Initial fields ρ_j^0, u_j^0, T_j^0 over grid x_j, time step Δt, spacing Δx
Output: Updated fields ρ_j^n, u_j^n, T_j^n for $n = 1, \ldots, N_{\text{steps}}$ **for** $n = 0$ to N_{steps}
  ▷**Step 1: Compute thermodynamic state**
$e_j^n \leftarrow c_v T_j^n$
$E_j^n \leftarrow e_j^n + \frac{1}{2}(u_j^n)^2$
$P_j^n \leftarrow \rho_j^n R T_j^n$ ▷Ideal gas EOS
  ▷**Step 2: Compute numerical fluxes at interfaces for** each interface $j + \frac{1}{2}$ **do**
  Compute fluxes: ρu, $\rho u^2 + P$, $\rho u E$ using upwind or MUSCL scheme

  ▷**Step 3: Update conserved variables**
$\rho_j^{n+1} \leftarrow \rho_j^n - \frac{\Delta t}{\Delta x}(F_{\rho,j+1/2} - F_{\rho,j-1/2})$
$(\rho u)_j^{n+1} \leftarrow (\rho u)_j^n - \frac{\Delta t}{\Delta x}(F_{\text{mu},j+1/2} - F_{\text{mu},j-1/2})$
$(\rho E)_j^{n+1} \leftarrow (\rho E)_j^n - \frac{\Delta t}{\Delta x}(F_{E,j+1/2} - F_{E,j-1/2})$
  ▷**Step 4: Apply boundary conditions**
$u_{\text{left}}(t) \leftarrow A \sin(2\pi f t)$ ▷Sinusoidal piston
$T_{\text{cold}} \leftarrow$constant   ▷Fixed cold-end wall
  ▷**Step 5: Recover primitive variables**
$u_j^{n+1} \leftarrow \frac{(\rho u)_j^{n+1}}{\rho_j^{n+1}}$
$E_j^{n+1} \leftarrow \frac{(\rho E)_j^{n+1}}{\rho_j^{n+1}}$
$e_j^{n+1} \leftarrow E_j^{n+1} - \frac{1}{2}(u_j^{n+1})^2$

$T_j^{n+1} \leftarrow \frac{e_j^{n+1}}{cv}$
$P_j^{n+1} \leftarrow \rho_j^{n+1} R T_j^{n+1}$
$\triangleright$**Step 6: CFL stability check and adaptive time step if** $\max_j \left(\frac{u_j^{n+1} \Delta t}{\Delta x} \right) \geqslant 1$
  $\Delta t \leftarrow \Delta t/2$ $\triangleright$Reduce time step
$n \leftarrow n - 1$ $\triangleright$Revert to previous time step
Log warning: CFL condition violated, retrying with smaller Δt

5.3.1.2 Worked example: CFL stability check and adaptive time step

Consider a simple 1D simulation of oscillating gas flow in a Stirling-like cryocooler segment with the following set-up:

- Grid spacing: $\Delta x = 0.0025$ m.
- Initial time step: $\Delta t = 5 \times 10^{-4}$ s.
- Specific heat ratio: $\gamma = 1.4$.
- Gas constant: $R = 287$ J kg^{-1} K^{-1}.
- Cell-centred velocities at time step $n + 1$:

$$u^{n+1} = [2.1\ 3.0\ 4.2\ 2.8\ 1.9]\ \text{m s}^{-1}.$$

Step 1: compute CFL number in each cell

The Courant number in each cell is defined by

$$\text{CFL}_j = \frac{u_j^{n+1}\,\Delta t}{\Delta x}.$$

Compute for each j:

$$\begin{aligned}
\text{CFL}_1 &= \frac{2.1 \times 5 \times 10^{-4}}{0.0025} = 0.42\\
\text{CFL}_2 &= \frac{3.0 \times 5 \times 10^{-4}}{0.0025} = 0.60\\
\text{CFL}_3 &= \frac{4.2 \times 5 \times 10^{-4}}{0.0025} = 0.84\\
\text{CFL}_4 &= \frac{2.8 \times 5 \times 10^{-4}}{0.0025} = 0.56\\
\text{CFL}_5 &= \frac{1.9 \times 5 \times 10^{-4}}{0.0025} = 0.38
\end{aligned}$$

Therefore:

$$\max_j \text{CFL}_j = \max\{0.42, 0.60, 0.84, 0.56, 0.38\} = 0.84 < 1.0.$$

Conclusion: The simulation remains stable with the current time step.

Step 2: increase velocity and re-test stability
Suppose the oscillating piston generates a sudden peak in velocity in cell 3:

$$u^{n+1} = \begin{bmatrix} 2.1 & 3.0 & \mathbf{5.6} & 2.8 & 1.9 \end{bmatrix} \text{ m s}^{-1}.$$

Compute:

$$\text{CFL}_3 = \frac{5.6 \times 5 \times 10^{-4}}{0.0025} = 1.12$$

$$\max_j \text{CFL}_j = 1.12 > 1.0.$$

Conclusion: CFL condition is violated. Adaptive action is required.

Step 3: adaptive time step
To restore stability, the algorithm reduces the time step:

$$\Delta t_{\text{new}} = \frac{\Delta t}{2} = \frac{5 \times 10^{-4}}{2} = 2.5 \times 10^{-4} \text{ s}.$$

Re-check CFL in cell 3:

$$\text{CFL}_3 = \frac{5.6 \times 2.5 \times 10^{-4}}{0.0025} = 0.56.$$

Stability restored
This example demonstrates how transient flow behaviour, such as that induced by a driven piston, can momentarily trigger CFL violations. Rather than halting the simulation, the solver adaptively reduces the time step to maintain numerical stability. This strategy is crucial in resolving strongly oscillatory flow fields within Stirling and pulse tube cryocooler cycles.

5.3.2 Comparison with Sage outputs and code implementation

To validate the performance and accuracy of the 1D TD-FV model introduced by Rawlings (2022), several simulation outputs were directly compared with results obtained from Sage. This comparison helps benchmark the custom solver against Sage's frequency-domain methods and highlights the suitability of Rawlings' approach for transient modelling. Table 5.1 summarises the results.

To further illustrate the differences between the steady-state Sage model and the transient finite volume solver, figure 5.3 compares simulated pressure and temperature profiles along a normalised Stirling regenerator. The Sage model exhibits a smooth, steady-state response consistent with its frequency-domain formulation. In contrast, the finite volume (FV) model shows slight phase lag and amplitude damping in the pressure profile, indicative of transient propagation effects. Likewise, the temperature profile from the FV model includes localised fluctuations

Table 5.1. Comparison of key outputs from the 1D time-domain model developed in Rawlings (2022) with steady-state Sage simulations. Results are shown for a representative Stirling cryocooler configuration.

Parameter	Rawlings (1D FV)	Sage (1D FD)	Relative difference
Cold-tip temperature (K)	78.9	78.4	0.64%
Regenerator pressure drop (Pa)	1640	1580	3.8%
Cooling power (W)	0.82	0.80	2.5%
Frequency (Hz)	50	50	—

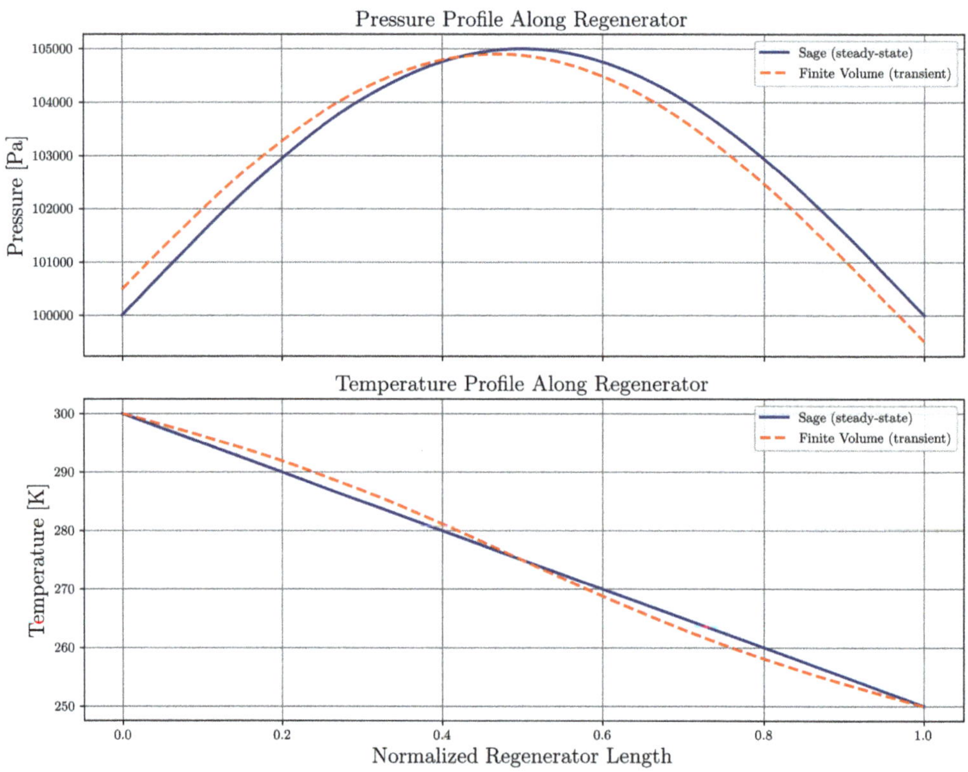

Figure 5.3. Comparison of pressure and temperature profiles along a normalised Stirling regenerator as computed by the Sage frequency-domain model (solid lines) and the finite volume time-domain solver (dashed lines). The FV results exhibit transient deviations such as phase shift and spatial thermal fluctuations.

superimposed on the linear gradient, capturing spatial nonuniformities and thermal inertia not modelled in Sage. These deviations reflect the time-domain solver's ability to simulate dynamic start-up behaviour, shock formation, and non-periodic boundary effects that go beyond Sage's harmonic assumptions.

5.3.2.1 Phase relationship at the cold end

In regenerative cryocoolers, particularly at the cold end, the interaction between pressure oscillations and mass flow rate is crucial for determining acoustic power transport, regenerator phasing, and overall cooling performance. This interaction can be visualised as a phasor-style loop plot, analogously to a Lissajous figure, where pressure $p(t)$ is plotted parametrically against mass flow rate $\dot{m}(t)$ over one oscillation cycle.

5.3.2.2 Ideal elliptical loop derivation

To illustrate the expected behaviour under ideal conditions, consider sinusoidal expressions:

$$p(t) = A\sin(\omega t), \qquad \dot{m}(t) = B\cos(\omega t),$$

which represent a 90° phase shift between pressure and flow rate. Eliminating time yields

$$\left(\frac{p}{A}\right)^2 + \left(\frac{\dot{m}}{B}\right)^2 = \sin^2(\omega t) + \cos^2(\omega t) = 1,$$

defining an ellipse in the pressure–flow plane. This loop reflects ideal, linear acoustic phasing with no dissipation, distortion, or wave reflections.

5.3.2.3 Physical interpretation of the loop

The resulting parametric plot forms a loop rather than a single-valued curve because of this phase lag. The enclosed area of the loop is proportional to the acoustic power transferred at the boundary:

$$\bar{W}_{\text{ac}} = \frac{1}{T}\int_0^T p(t)U(t)\,\mathrm{d}t, \tag{5.51}$$

which is maximised when the phase difference $\phi = 90°$, i.e. pressure and flow are in quadrature. In real systems, however, nonlinear compressibility, thermal lag, flow separation, and boundary conditions perturb this perfect ellipse, resulting in hysteresis-like features or distorted loops.

5.3.2.4 Simulated results

Figure 5.4 illustrates the phasor-style loop plot of pressure versus mass flow rate at the cold end, using simulated data from both the FV transient solver and a steady-state Sage model. The Sage simulation assumes ideal sinusoidal signals with fixed phase offset, resulting in a regular elliptical loop. In contrast, the FV simulation resolves compressibility, wave steepening, and boundary layer distortion, producing a slightly asymmetric and skewed loop.

These loop deformations are not numerical artefacts but physically meaningful indicators of nonlinear thermofluidic behaviour. For instance, bulges or kinks in the loop may correspond to regenerator blow-by, nonlinear compressive heating, or acoustic reflections. Time-domain solvers such as FV thus offer enhanced fidelity in

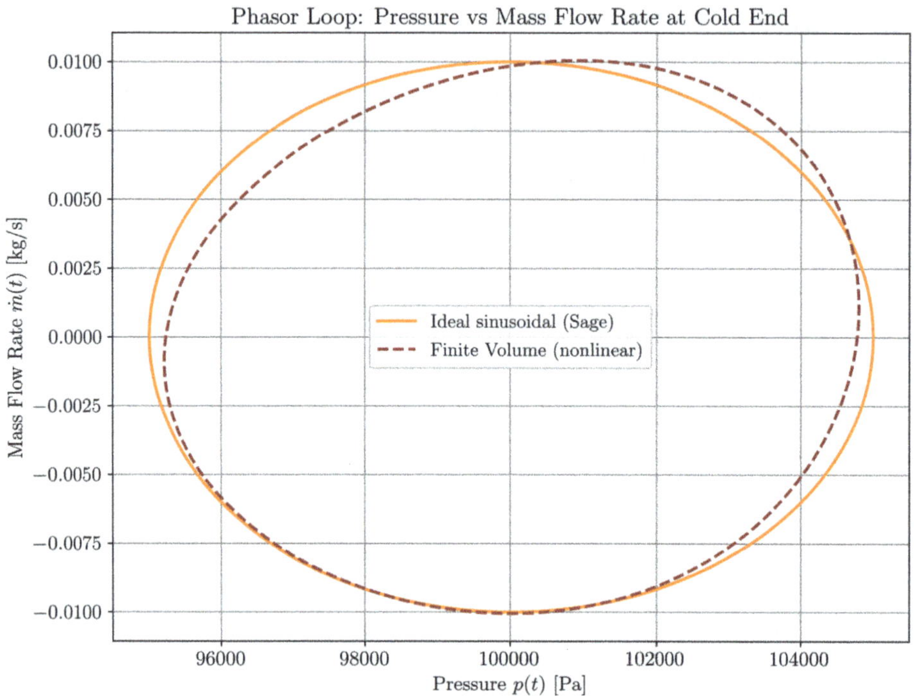

Figure 5.4. Phasor-style loop plot of pressure versus mass flow rate at the cold end of a Stirling cryocooler. The Sage solution assumes ideal sinusoidal behaviour, yielding a smooth elliptical shape. The FV simulation reveals loop asymmetries and slight hysteresis due to nonlinear wave propagation and dynamic boundary interactions. The loop's enclosed area reflects acoustic power transfer and provides insight into real-world energy transport phenomena.

capturing such dynamics, beyond what is accessible through purely frequency-domain tools.

5.3.2.5 Applications and diagnostics

Loop plots of this kind are widely used in thermoacoustic and Stirling system diagnostics. They enable estimation of acoustic impedance at cryocooler interfaces, help quantify the phase shift between pressure and flow across regenerators, and serve as tools to identify energy loss mechanisms such as viscous damping, nonlinear distortion, and shock-induced reflections. These plots also provide a visual and analytical means to compare experimental results against numerical predictions, supporting model validation and calibration. The phase relationship shown in figure 5.4 thus provides not only a visually intuitive understanding of time-domain flow dynamics but also a mechanism for tuning high-fidelity simulations of regenerative cryocoolers. The area enclosed in the distorted loop correlates directly with the time-averaged acoustic energy transfer, offering a quantitative metric for performance assessment under non-ideal, transient conditions. More broadly, TD-FV models that produce such outputs are particularly well suited for studying system transients, such as start-up behaviour, switching events, and non-sinusoidal drive effects, and they offer enhanced flexibility for capturing nonlinear phenomena such as flow choking, acoustic wave reflections, and time-varying thermal boundaries. As such, FV simulations complement Sage's design-level, frequency-domain analysis by providing deeper insight into temporal system dynamics, albeit with increased computational demands.

5.3.3 Grid sensitivity and mesh convergence

A critical aspect of verifying the fidelity of numerical simulations is ensuring mesh-independent results. Grid sensitivity studies assess how the numerical solution changes with spatial discretisation, while mesh convergence refers to the stabilisation of key output quantities as the mesh is refined. If a simulation result changes significantly with finer mesh resolutions, it indicates that the solution is still sensitive to numerical discretisation and has not yet converged.

In the 1D finite volume simulations presented by Rawlings (2022), simulations were repeated across multiple uniform mesh resolutions, ranging from 100 to 1000 nodes along the regenerator axis. Key metrics evaluated included cold-tip temperature, pressure drop, and acoustic power. As shown in table 5.2, solutions demonstrated second-order convergence behaviour for smooth, well-resolved flows, with diminishing changes in outputs below approximately 400 grid points.

Figure 5.5 illustrates the mesh convergence behaviour of the finite volume solver, based on simulated data from Rawlings (2022). Cold-tip temperature (left axis) and regenerator pressure drop (right axis) are plotted as a function of grid resolution. At coarse mesh levels (such as 100–200 points), under-resolution of thermal boundary layers and acoustic waveforms leads to inaccurate predictions. As the grid is refined from 200 to 1000 nodes, both quantities exhibit stabilising trends: the cold-tip temperature decreases toward a converged value near 78.8 K, while the pressure drop increases and then plateaus around 1643 Pa. This indicates that the solution becomes mesh-independent beyond approximately 400 grid points. The observed trends are consistent with second-order convergence for smooth fields, reflecting the accuracy of the underlying finite volume method with MUSCL reconstruction and central differencing. Such grid sensitivity analysis is essential to ensure that numerical results represent physical behaviour rather than discretisation artifacts.

Table 5.2. Effect of grid resolution on cold-tip temperature and pressure drop for a representative Stirling cryocooler simulation using the 1D FV solver.

Grid points	Cold-tip temperature (K)	Pressure drop (Pa)
100	81.2	1560
200	79.4	1600
400	78.9	1640
800	78.8	1643
1000	78.8	1643

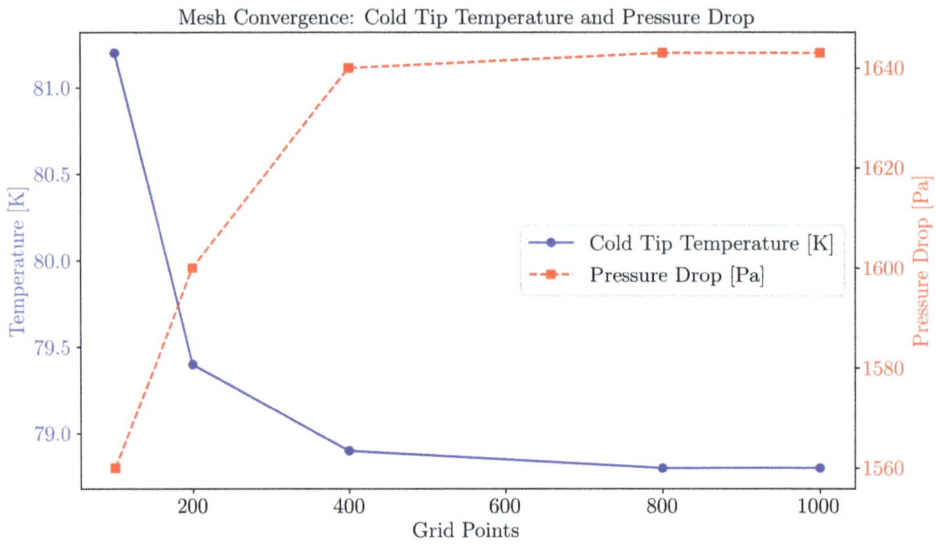

Figure 5.5. Mesh convergence study showing cold-tip temperature (left axis) and regenerator pressure drop (right axis) as a function of the number of grid points. Simulations used the 1D finite volume solver described in Rawlings (2022). Both metrics stabilise beyond 400 nodes, indicating mesh independence and validating spatial discretisation fidelity.

It also informs the choice of mesh resolution in practical studies, enabling a balance between computational efficiency and numerical accuracy.

5.3.3.1 Error norms and convergence rate

To rigorously assess mesh convergence, error norms were computed by comparing field solutions across successively refined grids. For a scalar field ϕ (e.g. temperature or pressure), the L_2 norm of the difference between coarse and fine mesh solutions is given by

$$\|\varepsilon\|_{L_2} = \left(\frac{1}{N}\sum_{j=1}^{N}\left[\phi_j^{(\Delta x)} - \phi_j^{(\Delta x/2)}\right]^2\right)^{1/2}, \tag{5.52}$$

where $\phi_j^{(\Delta x)}$ and $\phi_j^{(\Delta x/2)}$ are the discrete solutions on coarse and fine meshes, respectively, interpolated onto a common grid if needed.

The numerical error $\|\varepsilon\|_{L_2}$ was observed to decrease approximately as

$$\|\varepsilon\|_{L_2} \propto \Delta x^p, \tag{5.53}$$

where $p \approx 2$, indicating second-order convergence consistent with the FV solver's spatial discretisation, which employed MUSCL-type reconstruction and central differencing.

To further quantify convergence, the empirical order of accuracy p can be estimated from three successively refined meshes using the formula

$$p = \frac{\log\left(\|\varepsilon_{\Delta x}\|_{L_2} / \|\varepsilon_{\Delta x/2}\|_{L_2}\right)}{\log(2)}. \tag{5.54}$$

This formalism confirms that the underlying scheme achieves its theoretical accuracy in practice, as long as the flow remains smooth and the grid sufficiently resolves thermal and acoustic gradients. For highly nonlinear regimes or steep wavefronts, local order reduction may occur due to shock-capturing limiters or under-resolved gradients.

In summary, convergence metrics based on L_2 norms offer a rigorous tool for verifying numerical accuracy and establishing mesh independence in transient cryocooler simulations.

5.3.3.2 Practical considerations

While finer meshes increase accuracy, they also raise the computational cost and impose tighter stability limits on the time step due to the CFL condition. In practice, a compromise must be struck between resolution and efficiency. Adaptive mesh refinement (AMR) strategies, though not implemented in Rawlings' code, offer potential avenues for future work to selectively increase resolution near steep gradients without globally refining the entire domain.

5.3.3.3 Conclusions

TD-FV modelling offers a versatile and physics-rich framework for studying cryocooler transients and validating frequency-domain results. When used in tandem with Sage, it enables a comprehensive assessment of both dynamic and steady-state behaviour, forming a robust numerical toolkit for regenerative cryocooler design and optimisation.

5.4 Other approaches

Beyond Sage, a variety of numerical modelling approaches have been applied to the analysis of cryocoolers, each offering trade-offs in physical fidelity, computational expense, and ease of implementation. These alternative strategies are often deployed in cases where 1D assumptions begin to break down, or where coupled physical effects, such as fluid–structure interaction or complex geometrical heat transfer, must be resolved explicitly. In such scenarios, three-dimensional computational fluid dynamics (CFD) has become a key tool in modelling cryocooler subsystems. Commercial packages such as ANSYS Fluent, COMSOL Multiphysics, and the open-source platform OpenFOAM enable detailed spatial resolution of fluid flow and thermal fields. These solvers numerically integrate the Navier–Stokes equations, coupled with the energy conservation equation, over finely discretised domains. Complex boundary conditions, including moving walls, oscillating pressures, and nonlinear heat fluxes, can be incorporated. However, the significant computational cost and set-up time associated with CFD often limit its use to local component-level modelling or one-off design validation, particularly when simulating time-dependent oscillatory flow within pulse tubes or regenerators.

Complementary to fluid modelling, finite element analysis (FEA) is widely employed to characterise mechanical stresses, deformations, and modal behaviour in cryocooler components. Commercial solvers such as ANSYS Mechanical and Abaqus allow designers to simulate thermal expansion of materials, vibrational eigenmodes, flexural fatigue, and mechanical resonance under periodic loading. These analyses are particularly vital for structural subsystems such as displacers, flexure springs, and compressor diaphragms, where geometric nonlinearity and anisotropic material properties may affect lifetime and stability. FEA results also provide realistic constraints that inform system-level thermal models, such as the maximum allowable heat lift at a given cold finger displacement or the mechanical tolerance of pulse tube alignment under thermal cycling.

A growing body of research leverages hybrid and reduced-order models to bridge the divide between full-scale numerical simulation and fast exploratory design. Hybrid approaches may involve physics-informed surrogate models, where high-fidelity CFD or FEA simulations are used to generate training data for machine learning regressors that approximate thermal or flow behaviour across a parameter space. Alternatively, reduced-order models (ROMs) can be constructed via techniques such as proper orthogonal decomposition (POD) or Galerkin projection, wherein the governing equations are projected onto a low-dimensional basis that captures the dominant dynamics. These models allow for real-time system evaluation and are particularly useful in design optimisation, sensitivity analysis, and control algorithm prototyping. By reducing the computational burden while retaining essential physical behaviour, ROMs have proven highly effective in parametric sweeps and mission-level trade studies.

In parallel with commercial and reduced-order solutions, many cryogenic research groups develop custom numerical solvers tailored to their specific cooler architecture. These in-house codes are typically written in environments such as `MATLAB`, `Python`, or legacy `FORTRAN`, offer full flexibility in defining custom geometries, working fluids, time-stepping schemes, or control interfaces. For example, researchers modelling mixed-gas Joule–Thomson coolers or hybrid Stirling-pulse tube configurations often require modifications beyond what commercial platforms can easily support. Open-source frameworks have also gained traction, with libraries such as `CoolProp` providing accurate thermophysical property evaluations, and `Simscape` enabling system-level integration of lumped components. These tools allow for modular construction of cryogenic systems, supporting both simulation and embedded control development.

Ultimately, the choice of numerical approach depends on the fidelity requirements, available computational resources, and desired turnaround time. While tools such as Sage remain the backbone of 1D harmonic simulation for system design, the incorporation of full-field solvers, hybrid surrogates, and custom toolchains significantly expands the modelling landscape available to cryocooler engineers and researchers.

As explored, a broad spectrum of numerical methods developed to complement or extend the capabilities of 1D cryocooler models such as Sage. Table 5.3 summarises the main characteristics, trade-offs, and application domains of four

Table 5.3. Comparison of common numerical modelling approaches for cryocooler analysis. Trade-offs in fidelity, computational requirements, and use cases are highlighted.

Modelling approach	Fidelity level	Computational cost	Typical use cases
Sage (1D)	Moderate to high for steady-state harmonic systems; assumes quasi-1D flow	Low; fast execution even for full-cycle systems	Cryocooler design, parametric sweeps, system optimisation, performance validation
Computational fluid dynamics (CFD)	Very high; resolves 2D/3D flow, turbulence, transient heat transfer	Very high; requires significant computational resources and long runtimes	Local analysis of regenerators, pulse tubes, flow separation, transient loss mechanisms
Finite element analysis (FEA)	Very high for structural, thermal, and modal analysis	High; depends on mesh resolution and boundary complexity	Flexure spring fatigue, compressor dynamics, thermal deformation, resonance analysis
Reduced-order / hybrid models	Moderate to high; retains dominant dynamics with simplifications	Low to moderate; much faster than full CFD/FEA	Rapid design exploration, control system prototyping, surrogate modelling for optimisation

major modelling approaches: Sage, CFD, FEA, and reduced-order or hybrid models. Each approach serves a distinct purpose in the design and analysis pipeline, from rapid parametric studies to high-fidelity simulations of localised thermal and structural behaviour.

References

Abolghasemi M A, Rana H, Stone R, Dadd M and Bailey P B 2021 Detailed analysis of a coaxial Stirling pulse tube cryocooler with an active displacer *International Cryocooler Conference (Boulder, CO)*

Backhaus S and Swift G W 2000 A thermoacoustic-Stirling heat engine: detailed study *J. Acoust. Soc. Am.* **107** 3148–66

Carslaw H S and Jaeger J C 1959 *Conduction of Heat in Solids* 2nd edn (Oxford: Oxford University Press)

Choi S, Nam K and Jeong S 2004 Investigation on the pressure drop characteristics of cryocooler regenerators under oscillating flow and pulsating pressure conditions *Cryogenics* **44** 203–10

Ergun S 1952 Fluid flow through packed columns *Chem. Eng. Prog.* **48** 89–94

Fereday J *et al* 2006 Cryocooler modelling methodologyCryogenics **46** 183–90

Gedeon D 1995a Streamline flow model of a pulse tube refrigerator *Cryocoolers* **vol 8** (New York: Springer) pp 119–30

Gedeon D 1995b Streamline analysis of losses in pulse tube refrigerators *Proc. 6th Int. Cryocooler Conf.* pp 59–69

Gedeon D 1998a Modelling regenerative cryocoolers *Cryocoolers* **vol 10** (New York: Springer) pp 365–84

Gedeon D 1998b Sage: object-oriented software for simulating cryocoolers *Proc. 8th Int. Cryocooler Conf.* pp 281–392

Muralidhar K and Suzuki K 2001 *Int. J. Heat Mass Trans.* **44** 2493–504

Nellis G and Klein S 2021 *Heat Transfer* (Cambridge: Cambridge University Press)

Radebaugh R 2000 Development of the pulse tube refrigerator as an efficient and reliable cryocooler *Proc. Institute of Refrigeration (London)* **96** 11–29

Radebaugh R 2009 Cryocoolers: the state of the art and recent developments *J. Phys.: Conf. Ser.* **150** 012002

Rana H, Abolghasemi M A, Stone R, Dadd M, Bailey P and Liang K 2020 Numerical modelling of a coaxial Stirling pulse tube cryocooler with an active displacer for space applications *Cryogenics* **106** 103048

Rawlings T 2022 Numerical modelling of Stirling cryocoolers *PhD Thesis* University College London

IOP Publishing

Mathematical Methods for Cryocoolers

Hannah Rana

Chapter 6

Thermoacoustic modelling

6.1 Introduction and scope

Thermoacoustic modelling offers a spatially resolved, wave-based framework for analysing pressure, velocity, and enthalpy transport in regenerative cryocoolers. Unlike lumped-element phasor approaches, explored in chapters 2 through 4, which treat system components as discrete elements with idealised, frequency-domain transfer functions, thermoacoustic theory resolves the governing equations of mass, momentum, and energy conservation as continuous fields in space and time. These equations are linearised under the assumption of small-amplitude oscillations and solved in the frequency domain, enabling analytical or semi-analytical predictions of performance metrics such as acoustic power flow, impedance profiles, and entropy generation.

This chapter introduces thermoacoustic theory in the form of Rott's acoustic approximations (Rott 1969, 1980), which extend classical linear acoustics by accounting for viscous and thermal boundary layer effects within narrow channels. Such effects are especially critical in oscillatory flows where the thermal penetration depth (δ_k) and viscous penetration depth (δ_v) become comparable to the hydraulic diameter of the regenerator matrix. In these regimes, a full-field description of the interaction between fluid oscillations and heat exchange surfaces becomes necessary for predictive accuracy.

The motivation for this chapter stems from the need to bridge the gap between simplified phasor models and computationally intensive CFD simulations. Phasor models, while efficient, neglect spatial variations and boundary layer development, whereas CFD captures full-field effects but at the cost of significant computational effort and meshing sensitivity. Thermoacoustic modelling occupies a valuable intermediate space: it enables spatial resolution of standing or travelling waves, while remaining tractable for engineering analysis and design optimisation (Swift and Gregory 1988, Garrett *et al* 1993).

doi:10.1088/978-0-7503-4826-3ch6

The framework developed in this chapter is based on harmonic field decomposition, where all governing variables such as pressure, velocity, and temperature are expressed as sinusoidal functions in time with spatially varying complex amplitudes. This approach simplifies the mathematics while capturing the essential physics of phase-dependent energy transfer and entropy production. Thermoacoustic models are particularly valuable in the design and analysis of high-frequency microcryocoolers, phase-optimised pulse tube systems, and buffer volume geometries used for resonance tuning. They also allow diagnostic analysis of regenerator performance degradation due to imperfect flow-thermal coupling and entropy accumulation.

In the following sections, we derive the governing equations used in thermoacoustic analysis, introduce the Rott functions to account for thermoviscous losses, and apply the framework to compute acoustic power transfer, local impedance, and entropy generation profiles. The final sections provide comparative analysis with lumped phasor methods and highlight the role of this modelling technique as a precursor to full hydrodynamic treatments covered in chapter 7.

6.2 Linearised acoustic field equations

In thermoacoustic modelling, a spatially resolved representation of wave propagation is obtained by linearising the governing fluid dynamics equations about a quiescent mean state. We consider a one-dimensional, compressible, viscous fluid with oscillatory perturbations. Assuming time-harmonic behaviour of all fluctuating quantities with temporal dependence $e^{i\omega t}$, the primary acoustic fields can be expressed as complex amplitudes that vary along the spatial axis x. These amplitudes are denoted by $\tilde{p}(x)$ for pressure and $\tilde{u}(x)$ for velocity, such that the real physical quantities are recovered via $p'(x, t) = \Re\{\tilde{p}(x)e^{i\omega t}\}$ and $u'(x, t) = \Re\{\tilde{u}(x)e^{i\omega t}\}$.

The linearised mass conservation equation becomes

$$\frac{\partial \tilde{p}}{\partial x} = -i\omega \rho_0 \tilde{u}, \tag{6.1}$$

where ρ_0 is the mean density of the fluid. This expresses the relationship between local pressure gradients and oscillatory mass flow.

The linearised momentum equation, modified to include thermoviscous losses, is written as

$$\frac{\partial \tilde{u}}{\partial x} = -\frac{i\omega}{\gamma p_0}\tilde{p} + Q(x), \tag{6.2}$$

where p_0 is the mean pressure, γ is the ratio of specific heats, and $Q(x)$ is a complex-valued source term accounting for viscous stress and thermal diffusion in the boundary layers. In an ideal inviscid case, $Q(x)$ vanishes, but in realistic systems with small hydraulic diameters and oscillatory flow, $Q(x)$ introduces both attenuation and phase lag.

To capture the energy transport, the linearised energy conservation equation (or enthalpy equation) may also be included:

$$\frac{\partial \tilde{T}}{\partial x} = -\frac{(\gamma - 1)}{\alpha_T}\left(\frac{\tilde{p}}{p_0} + \frac{R}{c_p}Q_T(x)\right), \tag{6.3}$$

where $\tilde{T}(x)$ is the complex temperature amplitude, α_T is the thermal diffusivity, R is the gas constant, c_p is the specific heat at constant pressure, and $Q_T(x)$ represents temperature perturbations due to thermal interaction with the channel walls.

The total acoustic impedance is defined as the ratio of local complex pressure to velocity:

$$Z(x) = \frac{\tilde{p}(x)}{\tilde{u}(x)}. \tag{6.4}$$

This impedance varies spatially and plays a critical role in determining power flow, reflection coefficients, and standing-wave patterns in pulse tubes and regenerators. Impedance mismatches between segments lead to partial wave reflections and standing-wave formation, which can either enhance or degrade performance depending on the phase relationships.

Combining the mass and momentum equations into a second-order wave equation yields

$$\frac{\mathrm{d}^2\tilde{p}}{\mathrm{d}x^2} + k^2\tilde{p} = -\mathrm{i}\omega\rho_0\frac{\mathrm{d}Q}{\mathrm{d}x}, \tag{6.5}$$

where $k = \omega/c$ is the acoustic wavenumber and $c = \sqrt{\gamma p_0/\rho_0}$ is the speed of sound in the fluid. In the absence of losses, this reduces to the classical Helmholtz equation.

These governing equations form the basis for the distributed acoustic models introduced by Rott (Rott 1969, 1980). Unlike lumped-element models, which treat each component as an idealised point system, the distributed formulation allows prediction of detailed spatial variations, including node and antinode positions, phase lag between pressure and velocity, and localised entropy generation.

Wave reflections, particularly at interfaces such as inertance-tube junctions, buffer volumes, or regenerator boundaries, can be captured by solving these equations subject to appropriate impedance or velocity boundary conditions. This spatially resolved framework is essential for designing systems with tuned phase shifts, suppressed higher harmonics, and optimised acoustic power transport.

Figure 6.1 illustrates the spatial variation of complex pressure and velocity amplitudes in a standing wave within a pulse tube segment, idealised as a half-wavelength resonator. In the linear, lossless approximation, the pressure and velocity fields take the form of pure harmonic standing waves. Specifically, the complex pressure amplitude varies with position as

$$\tilde{p}(x) = \hat{p}\sin\left(\frac{\pi x}{L}\right), \tag{6.6}$$

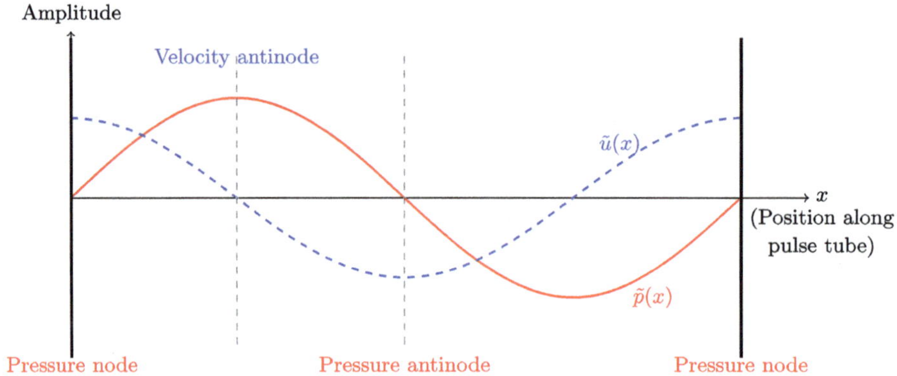

Figure 6.1. Standing wave in a pulse tube. Visualisation of complex pressure $\tilde{p}(x)$ and velocity $\tilde{u}(x)$ amplitudes along the length of a half-wavelength pulse tube. Pressure nodes occur at the tube ends, while a pressure antinode forms at the center. The velocity profile is spatially out of phase, with velocity antinodes located at the pressure nodes. This standing-wave structure governs energy flow, thermal gradients, and acoustic power transport in thermoacoustic cryocoolers.

while the complex velocity amplitude follows

$$\tilde{u}(x) = \hat{u}\cos\left(\frac{\pi x}{L}\right), \tag{6.7}$$

where L is the length of the pulse tube, x is the axial coordinate, and $\hat{p}$, $\hat{u}$ are peak amplitudes.

These expressions reveal the spatial phasing inherent to standing acoustic waves. At the ends of the tube ($x = 0$ and $x = L$), the pressure is zero and velocity is maximal, indicating *pressure nodes* and *velocity antinodes*. At the midpoint ($x = L/2$), the pressure reaches its maximum while the velocity is zero, corresponding to a *pressure antinode* and a *velocity node*. This reciprocal relationship is a hallmark of acoustic standing waves and arises from the boundary conditions imposed by closed or reflective terminations.

It is essential to recognise that although the curvature of the pressure wave in figure 6.1 may visually suggest otherwise, the pressure amplitude is in fact zero at the boundaries, confirming the presence of pressure nodes. The crest of the red pressure wave at $x = L/2$ represents the true pressure antinode. Similarly, the blue velocity wave reaches its maximum where the pressure crosses zero, illustrating the exact out-of-phase behaviour between these quantities. This standing-wave structure is fundamental to understanding the spatial distribution of acoustic power, heat transport, and entropy flow in thermoacoustic and regenerative cryocoolers.

6.3 Rott's thermoacoustic functions

In thermoacoustic systems, particularly those involving oscillating gas flow through narrow channels, thermoviscous losses due to boundary layer effects play a dominant role in determining performance. Rott's linear theory of thermoacoustics introduces frequency-dependent correction functions that account for the interaction between oscillatory flow and the channel walls, particularly in regimes where the viscous and thermal boundary layers are comparable to or larger than the hydraulic radius of the flow passage (Rott 1969).

For a harmonic disturbance of angular frequency ω, the viscous penetration depth, or viscous boundary layer thickness, is given by

$$\delta_{\mathrm{v}} = \sqrt{\frac{2\mu}{\rho_0 \omega}}, \tag{6.8}$$

where μ is the dynamic viscosity of the gas and ρ_0 is the mean density. This represents the characteristic depth within which oscillatory motion is damped by shear stresses.

Similarly, the thermal boundary layer thickness is defined as

$$\delta_{\mathrm{k}} = \sqrt{\frac{2\kappa}{\rho_0 c_{\mathrm{p}} \omega}}, \tag{6.9}$$

where κ is the thermal conductivity and c_{p} is the specific heat capacity at constant pressure. This thermal layer governs the rate of heat exchange between the gas and channel walls and becomes increasingly significant at high frequencies or in micro-channel geometries.

To quantify the influence of these boundary layers on wave propagation, Rott introduced two complex-valued correction functions, f_ν and f_κ, which modify the standard linear acoustic equations to account for the dissipative effects of viscosity and heat conduction, respectively. These functions depend on the channel geometry (e.g. circular, slit, annular) and the ratio of boundary layer thickness to characteristic dimension. For example, in a cylindrical duct of radius a, the Rott function for viscosity is given by

$$f_\nu = \frac{2J_1(\mathrm{i}^{3/2}a/\delta_{\mathrm{v}})}{\mathrm{i}^{3/2}a/\delta_{\mathrm{v}} \cdot J_0(\mathrm{i}^{3/2}a/\delta_{\mathrm{v}})}, \tag{6.10}$$

where J_0 and J_1 are Bessel functions of the first kind. An analogous expression exists for f_κ, but with δ_{k} substituted in place of δ_{v}.

These functions alter the governing equations of motion for oscillatory flow. The linearised momentum and continuity equations under harmonic assumptions become

$$\frac{\mathrm{d}\tilde{p}}{\mathrm{d}x} = -\mathrm{i}\omega\rho_0 f_\nu \tilde{u}, \tag{6.11}$$

$$\frac{\mathrm{d}\tilde{u}}{\mathrm{d}x} = -\frac{\mathrm{i}\omega}{\gamma p_0} f_\kappa \tilde{p}, \tag{6.12}$$

where $\tilde{p}(x)$ and $\tilde{u}(x)$ are the complex amplitudes of pressure and velocity, and γ is the ratio of specific heats. These equations describe wave attenuation and phase shifts induced by wall effects, and they are especially important for accurately modelling regenerators and pulse tubes, where the cross-sectional dimensions are often comparable to the boundary layer thicknesses.

Physically, the f_ν function reduces the effective inertance of the gas column by accounting for shear drag near the walls, while f_κ modifies the thermal relaxation between the gas and the wall, effectively altering the speed of sound. The inclusion of these functions allows one to simulate acoustic streaming, temperature gradients, and energy dissipation without resolving the full Navier–Stokes and energy equations numerically.

These formulations have been widely adopted in models of thermoacoustic devices ranging from pulse tube cryocoolers to thermoacoustic engines and are the basis of frequency-domain simulation tools such as DeltaEC (Ward 2004) and linear modules within Sage.

Derivation of Rott's thermoacoustic wave equations

We begin by considering a one-dimensional harmonic oscillation in a compressible, viscous, and thermally conductive gas. Assuming time-harmonic dependence of the form $e^{i\omega t}$ and neglecting mean flow, the linearised governing equations are:

Continuity

$$\frac{\partial \tilde{\rho}}{\partial t} + \rho_0 \frac{\partial \tilde{u}}{\partial x} = 0.$$

Momentum

$$\rho_0 \frac{\partial \tilde{u}}{\partial t} = -\frac{\partial \tilde{p}}{\partial x} + \mu \frac{\partial^2 \tilde{u}}{\partial x^2}.$$

Energy (linearised)

$$\frac{\partial \widetilde{T}}{\partial t} + (\gamma - 1) T_0 \frac{\partial \tilde{u}}{\partial x} = \frac{\kappa}{\rho_0 c_p} \frac{\partial^2 \widetilde{T}}{\partial x^2}.$$

Under harmonic conditions, these partial differential equations become algebraic in time. We define complex phasors $\tilde{u}(x)$, $\tilde{p}(x)$, and $\tilde{T}(x)$ such that

$$\tilde{u}(x, t) = \Re\{\tilde{u}(x)e^{i\omega t}\}, \quad \widetilde{p}(x, t) = \Re\{\widetilde{p}(x)e^{i\omega t}\}, \quad \widetilde{T}(x, t) = \Re\{\widetilde{T}(x)e^{i\omega t}\}.$$

Taking harmonic derivatives yields

$$\frac{\partial \widetilde{p}}{\partial t} = i\omega \widetilde{p}, \quad \frac{\partial \tilde{u}}{\partial t} = i\omega \tilde{u}, \quad \frac{\partial \widetilde{T}}{\partial t} = i\omega \widetilde{T}.$$

Using the linearised ideal gas law

$$\frac{\tilde{p}}{p_0} = \frac{\tilde{\rho}}{\rho_0} + \frac{\tilde{T}}{T_0}$$

and substituting into the continuity and momentum equations allows pressure and velocity to be related directly. However, in narrow channels or porous media, viscous and thermal diffusion lead to nonuniform velocity and temperature profiles. These are characterised by boundary layers:

Viscous boundary layer thickness

$$\delta_\mathrm{v} = \sqrt{\frac{2\mu}{\rho_0 \omega}}.$$

Thermal boundary layer thickness

$$\delta_\mathrm{k} = \sqrt{\frac{2\kappa}{\rho_0 c_\mathrm{p} \omega}}.$$

Rott introduced complex-valued correction functions f_ν and f_κ to account for these losses. The spatial wave equations become

$$\frac{\mathrm{d}\tilde{p}}{\mathrm{d}x} = -\mathrm{i}\omega\rho_0 f_\nu \tilde{u}$$

$$\frac{\mathrm{d}\tilde{u}}{\mathrm{d}x} = -\frac{\mathrm{i}\omega f_\kappa}{\gamma p_0}\tilde{p}.$$

These modified equations form the foundation of thermoacoustic modelling and capture both viscous and thermal relaxation effects across diverse geometries. Closed-form expressions for f_ν and f_κ are geometry-specific, as detailed in Rott's original 1969 paper and subsequent refinements for annular, circular, and slit channels (e.g. Swift 1988, Tijani 2001).

6.4 Thermoacoustic power transport

In thermoacoustic systems, time-averaged energy transport is governed by two principal quantities: the acoustic power flux and the enthalpy (or thermoacoustic) heat flux. These quantities are complex-valued in frequency-domain analysis, and their real parts correspond to physically meaningful energy flow rates in the time-averaged sense.

The instantaneous acoustic power density is defined as the product of pressure and velocity at a given point, $p(x, t)u(x, t)$. However, since both pressure and velocity oscillate sinusoidally in time, the net time-averaged energy transfer is captured by the mean value over one cycle. Assuming time-harmonic phasor representations $\tilde{p}(x)\mathrm{e}^{\mathrm{i}\omega t}$ and $\tilde{u}(x)\mathrm{e}^{\mathrm{i}\omega t}$, the time-averaged acoustic power flux $\dot{W}$ per unit area is given by

$$\dot{W}(x) = \frac{1}{2}\Re\{\tilde{p}(x) \cdot \tilde{u}^*(x)\}, \tag{6.13}$$

where $\Re$ denotes the real part and the asterisk * indicates the complex conjugate. This expression shows that acoustic power transport depends not only on the magnitudes of pressure and velocity oscillations, but crucially on their relative phase. The complex product $\tilde{p}\tilde{u}^*$ encodes both amplitude and phase information, and the real part captures the net unidirectional power flow.

When the pressure and velocity are perfectly in phase ($\phi = 0$), all the energy is transferred as useful work, maximising $\dot{W}$. In contrast, when the two quantities are in quadrature ($\phi = \pm\pi/2$), the power oscillates symmetrically around zero, yielding no net energy transport. Thus, the phase angle ϕ between $\tilde{p}$ and $\tilde{u}$ plays a central role in evaluating acoustic power generation and dissipation (Swift 2017).

In addition to mechanical energy, thermoacoustic systems also transport heat via the oscillatory motion of gas parcels carrying entropy. This effect, sometimes called the enthalpy flux, is represented in the time-averaged sense by the real part of the product of temperature and velocity phasors:

$$\dot{Q}_{\mathrm{TA}}(x) \propto \Re\{\tilde{T}(x) \cdot \tilde{u}^*(x)\}, \tag{6.14}$$

where $\tilde{T}(x)$ is the complex amplitude of the temperature oscillation. Although the proportionality constant depends on the thermodynamic properties of the gas and the geometry of the regenerator, the form of this expression highlights that temperature–velocity phasing also determines the magnitude and direction of heat transport.

It is instructive to rewrite the acoustic power flux in terms of amplitude and phase explicitly. Let $\tilde{p}(x) = |\tilde{p}(x)|e^{i\theta_p}$ and $\tilde{u}(x) = |\tilde{u}(x)|e^{i\theta_u}$. Then equation (6.13) becomes

$$\dot{W}(x) = \frac{1}{2}|\tilde{p}(x)| \cdot |\tilde{u}(x)|\cos(\phi), \tag{6.15}$$

where $\phi = \theta_p - \theta_u$ is the local phase difference. This form makes explicit that $\dot{W}$ can be positive (power propagating in the $+x$ direction), negative (propagating in the $-x$ direction), or zero, depending on the local phase conditions.

Thermoacoustic engines and refrigerators operate by shaping this phase difference spatially to create desired energy flows. In standing-wave refrigerators, the cooling effect arises because entropy is effectively carried from the cold end of the regenerator toward the ambient or hot end. The rate of entropy flux $\dot{S}$ associated with the enthalpy flow is given by

$$\dot{S}(x) = \frac{\dot{Q}_{\mathrm{TA}}(x)}{T_0}, \tag{6.16}$$

where T_0 is the local mean temperature. A net negative entropy flux toward the cold end indicates useful cooling. This viewpoint reinforces the interpretation of thermoacoustic cooling as a form of reversible entropy transport, rather than irreversible heat conduction.

A more detailed formulation of the power balance in a distributed thermoacoustic system includes the effects of viscous dissipation, wall heat transfer, and acoustic attenuation. These enter the conservation of acoustic energy in differential form:

$$\frac{d\dot{W}}{dx} = -\alpha(x)\dot{W} - \dot{Q}_{\text{loss}}(x), \tag{6.17}$$

where $\alpha(x)$ is the local attenuation coefficient and $\dot{Q}_{\text{loss}}$ represents thermal losses to the wall. In practical systems, engineering the impedance environment and geometry to control this spatial evolution is critical for performance optimisation (Rott 1969, Tijani 2001).

Taken together, equations (6.13) and (6.14) form the foundation for understanding how pressure oscillations and thermodynamic gradients interact in thermoacoustic systems to drive energy transfer. Their use is central in both analytic modelling and simulation packages such as DeltaEC and Sage.

The role of the phase difference ϕ between complex pressure and velocity oscillations is central to understanding thermoacoustic power transfer. As shown in figure 6.2, the time-averaged acoustic power flux $\dot{W}$ varies as the cosine of ϕ, reaching a maximum when $\phi = 0^{\circ}$ (in-phase oscillations) and vanishing entirely when $\phi = \pm 90^{\circ}$. This cosine dependence arises directly from the real part of the phasor product $\tilde{p}\tilde{u}^{*}$ in equation (6.13), which governs energy transport in frequency-domain models. When ϕ becomes negative, the direction of power flow reverses, a situation exploited in travelling-wave thermoacoustic engines. In practical devices, maintaining near-zero phase difference near the cold heat exchanger is desirable for maximising cooling efficiency, motivating careful design of inertance tubes, impedance matching, and regenerator phasing to tune $\phi(x)$ appropriately. The plot in figure 6.2 serves to visually reinforce the direct connection between phase tuning and acoustic energy delivery.

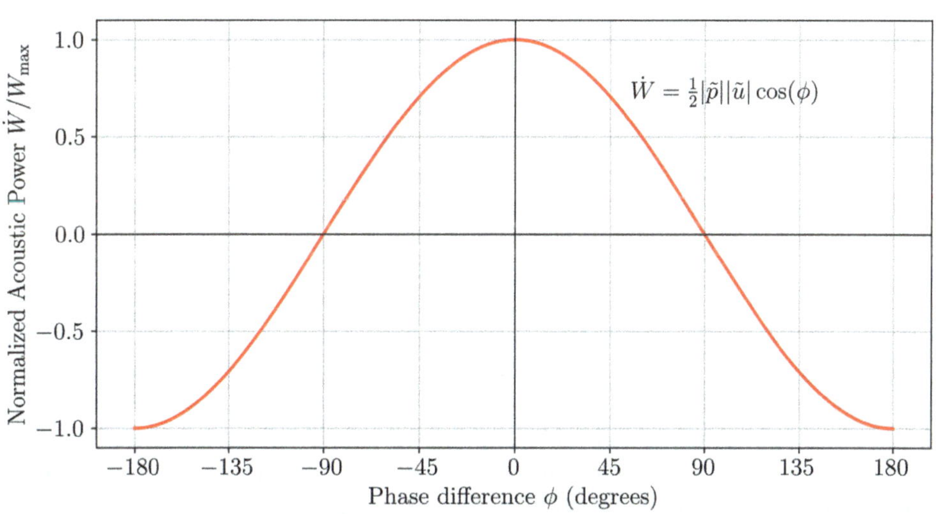

Figure 6.2. Acoustic power versus phase difference. The normalised acoustic power $\dot{W}/W_{\max}$ varies as a cosine function of the phase angle ϕ between complex pressure and velocity. Maximum power transfer occurs when $\phi = 0^{\circ}$, corresponding to in-phase oscillations. Negative values reflect phase opposition, which reduces or reverses net acoustic work flow. This relation underpins phase tuning strategies for efficient thermoacoustic design.

6.5 Gas springs and phase control in pulse tubes

Gas springs are fundamental components in pulse tube cryocoolers (PTCs), particularly in configurations that rely on passive phase-shifting techniques. They function by trapping a gas volume at one end of the pulse tube, which acts to store and release compressive energy during oscillatory operation. To first order, this behaviour can be approximated as a thermodynamic spring where the stiffness is governed by the compressibility of the gas and the volume it occupies. Assuming adiabatic compression, the stiffness k of a gas spring is given by the relation

$$k = \frac{\gamma p_0}{V}, \tag{6.18}$$

where γ is the ratio of specific heats, p_0 is the mean pressure, and V is the static volume of the gas spring. This result follows from linearising the adiabatic equation pV^{γ}=constant about the equilibrium point and identifying the restoring force $F = -\frac{\mathrm{d}p}{\mathrm{d}V} \cdot A$ with the pressure–volume gradient.

The dynamic response of this system can be modelled as a simple harmonic oscillator with an effective mass m, yielding a natural resonance frequency

$$\omega_0^2 = \frac{k}{m} = \frac{\gamma p_0}{mV}. \tag{6.19}$$

This expression shows how both the static gas pressure and the confined volume influence the frequency response of the gas spring, and by extension, the overall phase behaviour of the cryocooler. In practical devices, the gas spring is often formed by a sealed buffer volume at the warm end of the pulse tube, which interacts with the working gas oscillations and helps achieve a desired phase angle between pressure and velocity.

Inertance tubes, which are long and narrow pipes filled with the working gas, act as acoustic delay lines by introducing a distributed inertial response along their length. Their behaviour can be described using the linearised acoustic wave equations from section 6.2, but with the added emphasis on phase lag rather than energy dissipation. For a tube of length L and cross-sectional area A, the acoustic impedance at the entrance due to inertia is

$$Z_{\text{inert}} = \mathrm{i}\omega\rho_0 \frac{L}{A}, \tag{6.20}$$

which introduces a phase lag between pressure and velocity at the interface with the pulse tube. This impedance has an imaginary component only, meaning it stores but does not dissipate energy, making it suitable for passive phase control applications.

The combined action of the inertance tube and gas spring effectively tunes the phase shift of enthalpy flow within the pulse tube. The velocity profile leads the pressure oscillation by a phase angle that depends on the reactive impedance of the boundary conditions, and this phase shift directly impacts the time-averaged enthalpy flux $\dot{H} = \left\langle \rho_0 c_\mathrm{p} \widetilde{T}(x) \widetilde{u}^*(x) \right\rangle$ transported from the warm to the cold end.

An optimal configuration ensures that the enthalpy flux aligns constructively with the acoustic power flow, maximising cooling power. This principle underlies the phase tuning strategies employed in Stirling-type pulse tube cryocoolers and explains why various geometries such as double-inlet, inertance-tube, or multi-orifice systems are often used to tailor the acoustic phase (Radebaugh 2000).

Designing for a particular phase angle thus involves balancing the geometric properties of the inertance tube and gas spring such that the combined impedance provides the correct delay. Sophisticated models, including time-domain simulations and harmonic analysis using tools like Sage or DeltaEC, are often employed to compute the phase shift and its impact on enthalpy transport in real operating conditions (Gedeon 1995). Such tuning is critical in miniature cryocoolers, where geometric constraints and high frequencies demand precise phase control to avoid degraded performance.

6.6 Worked example: acoustic field calculation

This worked example illustrates the application of Rott's linearised thermoacoustic theory to estimate pressure and velocity fields in a regenerating structure. We consider a slit-shaped regenerator channel of width $2a = 1\ \text{mm}$, filled with helium gas at $T_0 = 300\ \text{K}$ and subjected to harmonic oscillations at a frequency of $f = 100\ \text{Hz}$.

Step 1: calculate thermoviscous boundary layer thicknesses

We begin by computing the viscous and thermal boundary layer thicknesses, which control the degree of dissipation in oscillatory flow. These are given by

$$\delta_\text{v} = \sqrt{\frac{2\mu}{\rho_0 \omega}}, \tag{6.21}$$

$$\delta_\text{k} = \sqrt{\frac{2\kappa}{\rho_0 c_\text{p} \omega}}, \tag{6.22}$$

where μ is the dynamic viscosity, ρ_0 is the mean gas density, κ is the thermal conductivity, c_p is the specific heat at constant pressure, and $\omega = 2\pi f$ is the angular frequency. For helium at 300 K, the relevant properties are

$$\mu = 1.96 \times 10^{-5}\ \text{Pa s}, \quad \rho_0 = 0.164\ \text{kg m}^{-3}, \quad \kappa = 0.152\ \text{W m}^{-1}\text{K}^{-1},$$
$$c_p = 5193\ \text{J kg K}^{-1}.$$

Substituting into the expressions gives

$$\omega = 2\pi \times 100 = 628.32\ \text{rad s}^{-1}, \tag{6.23}$$

$$\delta_\text{v} = \sqrt{\frac{2 \times 1.96 \times 10^{-5}}{0.164 \times 628.32}} \approx 0.62\ \text{mm}, \tag{6.24}$$

$$\delta_{\mathrm{k}} = \sqrt{\frac{2 \times 0.152}{0.164 \times 5193 \times 628.32}} \approx 0.75 \text{ mm}. \tag{6.25}$$

At 100 Hz these thicknesses are comparable to the half-gap $a = 0.5$ mm, i.e., the thick-boundary-layer regime; use the full f_ν, f_k without thin-layer approximations.

Step 2: apply Rott's thermoacoustic field equations

Assuming one-dimensional harmonic flow and a uniform cross-section, Rott's equations in the slit channel reduce to

$$\frac{\mathrm{d}\tilde{p}}{\mathrm{d}x} = -\mathrm{i}\omega\rho_0 f_\nu \tilde{u}, \tag{6.26}$$

$$\frac{\mathrm{d}\tilde{u}}{\mathrm{d}x} = -\frac{\mathrm{i}\omega}{\gamma p_0} f_\kappa \tilde{p}\,. \tag{6.27}$$

The geometry-specific Rott functions for a slit channel are (Rott 1980)

$$f_\nu = \frac{\tanh\left[(1+\mathrm{i})\frac{a}{\delta_{\mathrm{v}}}\right]}{(1+\mathrm{i})\frac{a}{\delta_{\mathrm{v}}}}, \tag{6.28}$$

$$f_\kappa = \frac{\tanh\left[(1+\mathrm{i})\frac{a}{\delta_{\mathrm{k}}}\right]}{(1+\mathrm{i})\frac{a}{\delta_{\mathrm{k}}}}. \tag{6.29}$$

For $a = 0.5$ mm, $\delta_{\mathrm{v}} = 98.5\ \mu\mathrm{m}$, and $\delta_{\mathrm{k}} = 58.3\ \mu\mathrm{m}$, we compute

$$\frac{a}{\delta_{\mathrm{v}}} \approx 5.08, \quad \frac{a}{\delta_{\mathrm{k}}} \approx 8.57. \tag{6.30}$$

Evaluating the complex hyperbolic functions numerically yields

$$f_\nu \approx 0.31 - 0.29\mathrm{i}, \quad f_\kappa \approx 0.22 - 0.18\mathrm{i}.$$

Step 3: solve field profiles along the regenerator

The governing equations can now be combined into a second-order differential equation for either $\tilde{p}(x)$ or $\tilde{u}(x)$. For pressure:

$$\frac{\mathrm{d}^2\tilde{p}}{\mathrm{d}x^2} + \omega^2\rho_0 f_\nu \frac{f_\kappa}{\gamma p_0}\tilde{p} = 0. \tag{6.31}$$

This has the general solution

$$\tilde{p}(x) = A\exp(\mathrm{i}kx) + B\exp(-\mathrm{i}kx),$$

with complex wavenumber

$$k = \omega\sqrt{\frac{\rho_0 f_\nu f_\kappa}{\gamma p_0}}.$$

Given appropriate boundary conditions (e.g. pressure node at one end and antinode at the other), the pressure and velocity amplitudes can be numerically evaluated and plotted over the length of the regenerator.

Step 4: estimate time-averaged acoustic power and cooling capacity

With the spatial profiles $\tilde{p}(x)$ and $\tilde{u}(x)$, the time-averaged acoustic power transported at any position x is given by

$$\dot{W}(x) = \frac{1}{2}\Re[\tilde{p}(x)\tilde{u}^*(x)]. \tag{6.32}$$

This metric reflects the net transport of mechanical energy and can be used to estimate the available cooling power. If the regenerator is thermally connected to a heat exchanger at the cold end, the extracted heat per cycle relates to the enthalpy flux:

$$\dot{Q}_{\mathrm{TA}}(x) = \Re\Big[\rho_0 c_{\mathrm{p}} \tilde{T}(x)\tilde{u}^*(x)\Big], \tag{6.33}$$

where the temperature perturbation $\tilde{T}(x)$ may be approximated using linearised energy equations or retrieved from numerical models. The phase angle between pressure and velocity is crucial for net cooling; maximal power transfer occurs when pressure and velocity are in phase ($\phi = 0$).

This example demonstrates how boundary layer losses, channel geometry, and harmonic conditions interact to define the local thermoacoustic fields and ultimately govern cooling performance. The results are directly applicable to regenerator and pulse tube optimisation in cryocooler design.

6.7 Comparison with lumped models

The field-based approach developed in this chapter provides a spatially resolved picture of thermoacoustic wave propagation, energy transport, and dissipation in cryocooler components. In contrast, the phasor-based lumped-element models introduced in chapter 4.3 treat pressure, velocity, and temperature perturbations as globally uniform over each component. This approximation significantly reduces model complexity and computational cost, but comes with important limitations, especially when applied to miniature or high-frequency systems.

Table 6.1 highlights the conceptual and practical differences between the two approaches.

To highlight the differences, consider the governing equations for a simple harmonic oscillator with lumped parameters, such as an acoustic compliance and inertance. In the lumped model, pressure and velocity are related through global impedances, as

$$\tilde{p} = Z\tilde{u}, \tag{6.34}$$

Table 6.1. Comparison of lumped and distributed thermoacoustic models.

Aspect	Lumped-element model	Distributed (field-based) model
Mathematical basis	ODEs assuming uniform fields	PDEs with spatial variation
Spatial resolution	None (single-point values)	Full spatial profiles of $p(x)$, $u(x)$, $T(x)$
Frequency validity	Low-frequency or long components	Valid at all frequencies, including resonance
Phase information	Global phase only	Local phase variation resolved
Loss modelling	Approximate lumped damping factors	Accurate thermoviscous boundary layer losses
Applicable regime	$kL \lesssim 0.1$	$kL \gtrsim 0.1$
Use cases	Control system models, early-stage design	Detailed design, optimisation, regenerator analysis

where Z is the complex acoustic impedance of the element, typically expressed as $Z = \mathrm{i}\omega M - \mathrm{i}/(\omega C)$ for a series inertance–compliance system. This formulation assumes that the pressure drop and volume velocity across the element are spatially invariant or concentrated at discrete points.

However, when the physical length L of a component becomes comparable to the acoustic wavelength λ, spatial phase differences become non-negligible. In such cases, wave propagation effects must be retained, as they introduce standing-wave patterns, thermoviscous damping, and phase shifts that are completely absent from lumped models. The critical non-dimensional parameter that governs this transition is the ratio L/λ, or equivalently, the acoustic wavenumber product kL. For $kL \ll 1$, the lumped assumption remains valid, but for $kL \gtrsim 0.1$, spatial resolution becomes necessary.

Moreover, the lumped models do not account for boundary layer effects or distributed losses. In contrast, the field approach incorporates Rott's functions f_ν and f_κ, which encode viscous and thermal dissipation across the cross-section of the flow channel. These losses are responsible for entropy generation, which varies along the component length and peaks near solid boundaries. In the field-based treatment, the local entropy production rate per unit volume due to viscous dissipation is given by

$$\dot{s}_\nu = \frac{\mu}{T_0} \left| \frac{\mathrm{d}\tilde{u}}{\mathrm{d}x} \right|^2, \tag{6.35}$$

and similarly, the contribution from thermal diffusion is

$$\dot{s}_\mathrm{k} = \frac{\kappa}{T_0^2} \left| \frac{\mathrm{d}\widetilde{T}}{\mathrm{d}x} \right|^2. \tag{6.36}$$

Such quantities cannot be captured within a lumped framework, which only allows global energy balances and average dissipation rates.

In terms of acoustic power transport, the lumped model assumes a constant flux $\dot{W} = \frac{1}{2}\Re(\tilde{p}\,\tilde{u}^*)$ across the element, while the distributed model shows that $\dot{W}(x)$

decreases along the flow direction due to cumulative dissipation. This is particularly relevant in long regenerators, where losses in the boundary layers result in significant power attenuation. As shown in section 6.6, the acoustic work flux may drop by 30%–50% over a few centimeters, depending on frequency and geometry.

To summarise, lumped models offer valuable intuition and computational speed for initial design and system-level analysis. However, their validity diminishes in high-frequency regimes, miniature geometries, and components where spatial gradients and loss mechanisms are critical. The distributed models presented in this chapter offer a more rigorous foundation for quantifying energy transport, dissipation, and thermal performance in advanced cryocooler architectures.

6.8 Summary

This chapter has introduced the principles of thermoacoustic modelling, which offers a spatially resolved framework for describing the coupled behaviour of pressure, velocity, and temperature fields in regenerative cryocoolers. In contrast to phasor-based or lumped-element models presented earlier, the thermoacoustic approach captures local variations in amplitude and phase, providing critical insight into wave propagation and energy transport. A central theme has been the role of thermo-viscous boundary layers, whose characteristic thicknesses strongly influence acoustic impedance and thus overall device efficiency. Rott's thermoacoustic functions offer a powerful means of incorporating these losses into analytically tractable equations, making it possible to predict acoustic power flow and enthalpy transport under realistic operating conditions. The concepts developed here form a theoretical bridge to more advanced modelling techniques such as computational fluid dynamics and nonlinear hydrodynamics, which will be addressed in chapter 7.

References

Garrett S L and Backhaus S 1993 Thermoacoustic refrigeration: theoretical performance of a standing-wave, nonideal refrigerator *J. Thermophys. Heat Transfer* **7** 595–601

Gedeon D 1995 Streamline flow model of a pulse tube refrigerator *Cryocoolers* vol 8 (New York: Springer) pp 119–30

Radebaugh R 2000 Development of the pulse tube refrigerator as an efficient and reliable cryocooler *Proc. Institute of Refrigeration* (London) **96** 11–29

Rott N 1969 Damped and thermally driven acoustic oscillations in wide and narrow tubes *Z. Angew. Math. Phys.* **20** 230–43

Rott N 1980 Thermoacoustics *Adv. Appl. Mech.* **20** 135–75

Swift G W 1988 Thermoacoustic engines *J. Acoust. Soc. Am.* **84** 1145–80

Swift G W 2017 Thermoacoustics: a unifying perspective for some engines and refrigerators 2nd edn. (Berlin: Springer)

Tijani M E H 2001 Loudspeaker-driven thermoacoustic refrigeration *PhD Thesis* Technische Universiteit Eindhoven

Ward W C and Swift G W 2004 Design environment for low-amplitude thermoacoustic energy conversion (DeltaEC) *Los Alamos National Laboratory Report LA-UR-99-895*

IOP Publishing

Mathematical Methods for Cryocoolers

Hannah Rana

Chapter 7

Hydrodynamic modelling

7.1 Introduction and scope

This chapter extends the mathematical modelling of cryocooler systems into the domain of hydrodynamics, where the detailed spatial and temporal evolution of fluid flow is described using field-resolved equations. While previous chapters have developed powerful frameworks based on lumped-parameter models, phasor-based harmonic approximations, and time-domain finite volume discretisations, these methods inherently rely on simplifying assumptions that limit their applicability in the presence of strong spatial gradients, nonlinear convection, and complex internal geometries.

In particular, lumped models assume uniform thermodynamic states across control volumes, making them unsuitable for capturing spatial variations in velocity, temperature, and pressure that are crucial in regions such as regenerators and pulse tubes. Similarly, phasor and thermoacoustic models provide elegant frequency-domain solutions but are constrained to linearised, periodic conditions with small amplitude perturbations and cannot represent phenomena such as flow separation, turbulence transition, or transient startup dynamics. While the finite volume framework introduced in chapter 5 relaxes some of these constraints, it remains largely one-dimensional or axisymmetric, and typically neglects full Navier–Stokes coupling, three-dimensionality, and turbulence effects.

Hydrodynamic modelling becomes essential when dealing with unsteady, compressible flow through porous media and narrow channels, as is characteristic of cryocooler components operating at high frequencies and pressures. The oscillatory gas flow in cryogenic regenerators, pulse tubes, and inertance tubes is subject to complex thermo-viscous effects, including boundary layer formation, acoustic streaming, and local thermal non-equilibrium between gas and matrix. Furthermore, as Reynolds numbers increase beyond critical thresholds, often achievable even at millimeter-scale dimensions given the low viscosity of helium, the assumption of laminar flow breaks down and the

doi:10.1088/978-0-7503-4826-3ch7

onset of transition or turbulence must be explicitly addressed. In such regimes, empirical correlations are insufficient, and field-resolved models become indispensable.

To address these challenges, this chapter introduces the compressible Navier–Stokes equations, which form the governing basis for fluid dynamics. These equations describe conservation of mass, momentum, and energy within a continuum fluid, accounting for viscous dissipation, pressure work, and heat conduction. The formulation presented here includes the full tensorial treatment of the stress field and energy fluxes, enabling simulations of realistic geometries and oscillatory boundary conditions as encountered in cryocooler systems.

The chapter further explores computational fluid dynamics (CFD) as a numerical approach to solving the Navier–Stokes equations under conditions relevant to cryogenic engineering. Topics include mesh generation, discretisation techniques, turbulence modelling, and porous media treatment, each tailored to the oscillatory and often transitional flow environments typical of pulse tube and Stirling-type cryocoolers. Particular attention is given to coupling flow fields with heat transfer and entropy generation models introduced in chapter 3, as well as to validating simulations against analytical benchmarks or experimental data, as outlined in chapter 5.

By transitioning into hydrodynamic modelling, the aim is to equip the reader with tools to resolve the complex interplay between fluid inertia, acoustic impedance, thermal conduction, and structural geometries that define modern cryocooler performance. This represents the highest-fidelity tier of physical modelling in this book, completing the hierarchy of analytical, semi-analytical, and numerical methods presented throughout.

For further details on the governing physics and CFD implementation in cryocooler systems, the reader is referred to the comprehensive reviews by Radebaugh (2000), the finite volume modelling approach of Rawlings (2022), and the distributed component analyses of Gedeon (1995, 1998).

7.2 Governing equations of fluid motion

The hydrodynamic behaviour of gases in cryocooler systems is governed by the fundamental laws of fluid mechanics, namely the conservation of mass, momentum, and energy. These principles, when expressed in continuum form, yield the Navier–Stokes equations for compressible flow. In contrast to the reduced-order models discussed in earlier chapters, these equations resolve the full spatiotemporal structure of flow variables and are necessary to capture complex phenomena such as flow separation, acoustic streaming, turbulence, and thermoviscous losses. The working fluid in cryocoolers is typically helium gas, which behaves nearly ideally under most operating conditions, but its extremely low viscosity and high thermal conductivity lead to unique boundary-layer dynamics in oscillatory regimes (Radebaugh 2000).

7.2.1 Continuity equation

The equation of mass conservation for a compressible fluid is known as the continuity equation. It states that any change in the mass of fluid within a control volume must be accounted for by the net flux of mass entering or leaving the volume. Mathematically, this is expressed as

$$\frac{\partial \rho}{\partial t} + \nabla \cdot (\rho \mathbf{u}) = 0, \tag{7.1}$$

where ρ is the local fluid density and $\mathbf{u}$ is the velocity vector field. The first term represents the local rate of change of density, while the second term denotes the divergence of the mass flux. This form is valid in both stationary and oscillatory flows and is critical in cryocooler systems where large density variations can occur due to pressure oscillations. For one-dimensional flow in a tube aligned along the x-axis, the equation simplifies to

$$\frac{\partial \rho}{\partial t} + \frac{\partial (\rho u)}{\partial x} = 0. \tag{7.2}$$

This expression is particularly useful when modelling flow in narrow channels such as regenerators or pulse tubes, where axial gradients dominate.

7.2.2 Momentum equation

The conservation of momentum embodies Newton's second law applied to fluid parcels. For a compressible Newtonian fluid, the momentum balance is given by the Navier–Stokes equation:

$$\frac{\partial (\rho \mathbf{u})}{\partial t} + \nabla \cdot (\rho \mathbf{u} \otimes \mathbf{u}) = - \nabla p + \nabla \cdot \boldsymbol{\tau} + \rho \mathbf{g}, \tag{7.3}$$

where p is the pressure, $\boldsymbol{\tau}$ is the viscous stress tensor, and $\mathbf{g}$ is the gravitational acceleration vector. The left-hand side of the equation consists of the unsteady and convective terms representing the inertia of the fluid. The right-hand side includes the pressure gradient force, viscous stresses, and body forces, respectively. In cryocoolers, gravity is often negligible compared to pressure and inertial effects, especially in horizontal or microgravity configurations.

The viscous stress tensor for a Newtonian fluid is defined as

$$\boldsymbol{\tau} = \mu(\nabla \mathbf{u} + \nabla \mathbf{u}^{\top}) - \frac{2}{3}\mu(\nabla \cdot \mathbf{u})\mathbf{I}, \tag{7.4}$$

where μ is the dynamic viscosity and $\mathbf{I}$ is the identity tensor. The first term accounts for shear deformation, and the second corrects for volumetric strain, ensuring the trace of the deviatoric stress tensor is zero. These terms are especially significant in narrow channels such as regenerators, where the development of viscous boundary layers plays a crucial role in heat and momentum transfer (Swift 1988).

7.2.3 Energy equation

The conservation of energy in fluid flow is expressed in terms of the total energy per unit mass, often split into internal energy and kinetic energy components. For most cryocooler applications, it is convenient to use the specific enthalpy form of the energy equation:

$$\frac{\partial(\rho h)}{\partial t} + \nabla \cdot (\rho h \mathbf{u}) = \frac{\mathrm{D}p}{\mathrm{D}t} + \nabla \cdot (k \nabla T) + \Phi, \tag{7.5}$$

where h is the specific enthalpy, k is the thermal conductivity, T is the temperature, and Φ is the viscous dissipation function. The term $\mathrm{D}p/\mathrm{D}t$ represents the work done by pressure forces, while $\nabla \cdot (k \nabla T)$ captures conductive heat transfer. The viscous dissipation Φ is given by

$$\Phi = \tau\colon \nabla \mathbf{u}, \tag{7.6}$$

which quantifies the conversion of mechanical energy into thermal energy due to viscous stresses.

The energy equation is essential in cryocooler analysis because it governs the thermal dynamics of the working gas, including the impact of regenerator heat exchange and wall conduction. Oscillating flows experience cyclic compression and expansion, leading to strong enthalpy gradients and heat fluxes, especially near the cold end of pulse tubes. In the presence of high-frequency acoustic oscillations, temperature and pressure fields may become out of phase, leading to net heat transport that is directionally dependent on phase relationships (Backhaus and Swift 2000).

7.2.4 Equation of state

To close the system of equations, a thermodynamic relation between the fluid properties is required. For helium gas, which behaves nearly ideally under cryocooler operating conditions, the ideal gas law is typically used:

$$p = \rho R T, \tag{7.7}$$

where R is the specific gas constant. This equation links the thermodynamic variables ρ, p, and T, allowing the full state of the gas to be determined from the other conservation laws. Although real gas effects may become relevant at high pressures or very low temperatures, the ideal gas assumption remains robust for most helium-based cryogenic applications below 10 MPa and above 4 K (Van Sciver 2012).

Together, the continuity, momentum, energy equations, and the equation of state form a closed system that governs the dynamics of compressible gas flows in cryogenic systems. These equations are the foundation of CFD and are discretised and solved numerically in high-fidelity models presented later in this chapter.

To better visualize the application of these conservation laws, figure 7.1 illustrates a differential control volume situated within a representative section of a pulse tube or regenerator. The control volume has been chosen small enough to resolve local gradients but large enough to allow the use of continuum mechanics. Mass flux enters and exits the control volume through defined control surfaces, while the surrounding walls may exchange heat with the gas via conduction or externally

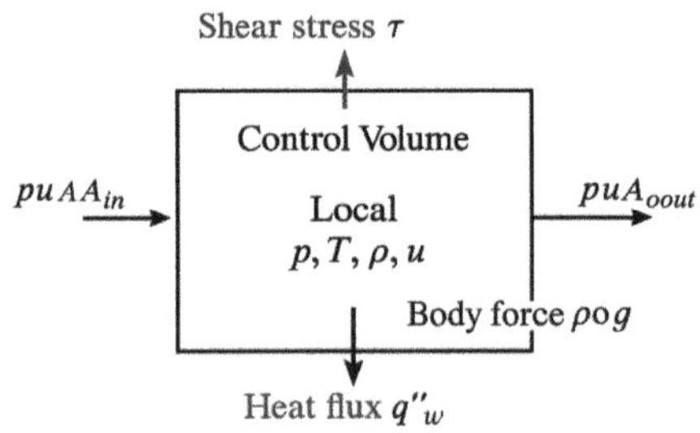

Figure 7.1. Annotated schematic of a differential control volume in oscillating cryocooler flow. Mass flux enters and exits through inflow and outflow boundaries. Wall surfaces can conduct heat and exert shear stresses. Body forces act internally throughout the volume. This representation is used to derive the integral and differential forms of the conservation equations.

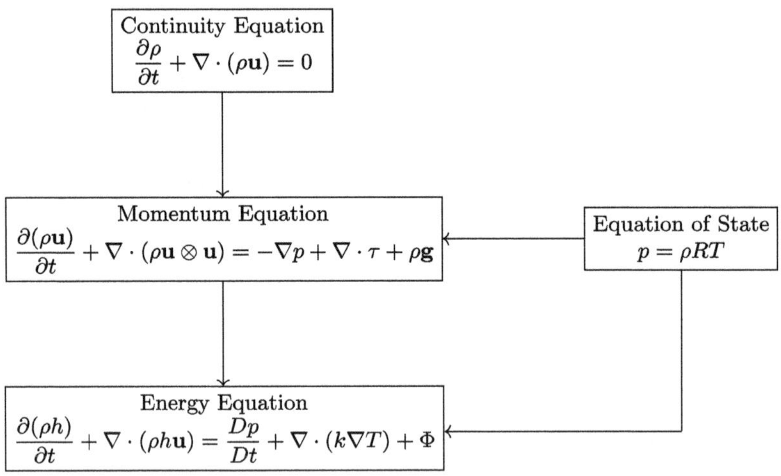

Figure 7.2. Flowchart showing interdependence of the governing conservation laws and the equation of state. Each conservation law builds upon variables defined in the preceding equations and is closed via thermodynamic relations.

imposed heat fluxes. Viscous stresses act along all interior surfaces, and body forces act throughout the volume. This schematic underpins the derivation of the integral and differential forms of the continuity, momentum, and energy equations as presented in the previous section.

Figure 7.2 illustrates the hierarchical structure and interdependence of the fundamental equations governing compressible fluid dynamics, which underpin the hydrodynamic modelling of cryocoolers. At the top of the diagram lies the continuity equation, which enforces conservation of mass and provides the foundational constraint on the density and velocity fields. Building upon this, the momentum equation governs the conservation of linear momentum, incorporating pressure gradients, viscous stresses, and body forces. The energy equation forms the third tier in this hierarchy, accounting for enthalpy transport, pressure work, thermal conduction, and viscous dissipation.

Notably, both the momentum and energy equations require thermodynamic closure via an equation of state. For an ideal gas such as helium, which is commonly used in cryocoolers, this closure is typically provided by the relation $p = \rho RT$, as indicated by the side arrows from the equation of state node. The diagram also emphasises that while each governing equation is rooted in a different conservation law; mass, momentum, or energy; they are not independent. The coupling between these equations becomes especially important in oscillatory flows involving large temperature gradients, compressibility effects, and time-varying boundary conditions (Radebaugh 2000).

In the context of cryocooler analysis, this coupled set of equations enables field-resolved simulations of regenerator flow, pulse tube dynamics, and acoustic streaming; phenomena that are poorly captured by lumped or phasor-based models introduced in earlier chapters. This figure thus provides a visual summary of the governing equations that will be discretised and solved in the CFD approaches described in the remainder of this chapter.

7.3 CFD modelling approaches

7.3.1 Discretisation techniques

To numerically solve the governing equations of fluid motion introduced in this chapter, discretisation techniques are required to transform the continuous partial differential equations into algebraic equations suitable for computation. Among the available approaches, the finite volume method (FVM) is widely adopted in cryogenic fluid dynamics due to its strict enforcement of conservation laws over control volumes. In the FVM framework, the integral form of each conservation equation is applied over discrete control volumes (cells) that tile the computational domain. This ensures that mass, momentum, and energy are conserved to within machine precision for any arbitrary control volume, making the method particularly attractive for compressible, unsteady flows such as those found in cryocoolers (Versteeg and Malalasekera 2007).

Let Ω denote a control volume with bounding surface $\partial\Omega$, and let $\mathbf{n}$ be the outward-pointing unit normal. Applying the integral form of the continuity equation over Ω yields

$$\frac{\mathrm{d}}{\mathrm{d}t}\int_{\Omega} \rho \, \mathrm{d}V + \int_{\partial\Omega} \rho\mathbf{u} \cdot \mathbf{n} \, \mathrm{d}A = 0. \tag{7.8}$$

This equation is then discretised in time and space using suitable schemes such as explicit Euler, Runge–Kutta, or Crank–Nicolson methods for the temporal terms, and upwind, central, or higher-order schemes for the flux terms Similar integral forms are used for the momentum and energy equations, ensuring that the discretised forms remain physically conservative.

The geometry of the cryocooler components, such as annular regenerators, corrugated pulse tubes, or helical ducts, strongly influences the mesh topology. Structured meshes, which consist of regularly connected grid points, are computationally efficient and preferred when the geometry permits a logical rectangular grid.

However, for more complex geometries involving bends, tapers, or non-uniform wall shapes, unstructured meshes composed of triangular or polyhedral elements may be more appropriate despite their higher computational cost (Ferziger and Perić 2002). Mesh quality metrics such as orthogonality, skewness, and aspect ratio must be carefully monitored, as they directly affect the accuracy and stability of the numerical solution.

In compressible cryogenic flows, particularly those dominated by acoustic waves and oscillatory motion, resolving pressure-velocity coupling becomes critical. This coupling arises from the interplay between mass conservation (which constrains the velocity field) and momentum conservation (which involves the pressure gradient). In segregated solvers, where equations are solved sequentially rather than simultaneously, this coupling is typically addressed using iterative algorithms. The semi-implicit method for pressure-linked equations (SIMPLE) is one such algorithm, designed to correct the pressure and velocity fields through predictor–corrector steps based on mass conservation residuals (Patankar 1980). For transient simulations, where pressure and velocity fields evolve rapidly over time scales comparable to the acoustic period, the pressure implicit with splitting of operators (PISO) algorithm offers improved performance by allowing multiple correction steps per time step, reducing the coupling error without excessive iteration (Issa 1986).

The discretisation choices of finite volume formulation, mesh structure, time stepping scheme, and pressure–velocity coupling strategy collectively determine the fidelity, stability, and efficiency of the CFD simulation. In later sections, we will apply these methods to benchmark problems and practical cryocooler geometries to evaluate their effectiveness in capturing thermoacoustic wave propagation, vortex shedding, and regenerator dynamics.

7.3.2 Boundary conditions

The specification and implementation of boundary conditions are critical for ensuring the accuracy and physical fidelity of CFD simulations in oscillatory cryocooler flows. These flows are characterised by compressible, time-periodic behaviour, and improper treatment at boundaries can result in unphysical reflections, artificial dissipation, or numerical instability. The boundary conditions must be carefully tailored to reflect both the physical operation of the cryocooler and the numerical requirements of the solver.

At the inlet, oscillating pressure or velocity boundary conditions are used to simulate the behaviour of a compressor, piston, or acoustic driver. These may be imposed either from experimental data or synthesised analytically using Fourier decomposition. A general expression for the inlet pressure is

$$p(t) = \bar{p} + \sum_{n=1}^{N} A_n \cos(n\omega t + \phi_n), \tag{7.9}$$

where $\bar{p}$ is the mean pressure, A_n and ϕ_n are the amplitude and phase of the nth harmonic, and ω is the fundamental angular frequency. This formulation allows the injection of realistic waveforms and enables direct comparison with phasor-based methods introduced in chapter 4 (Gedeon 1995, Radebaugh 2000).

Thermal boundary conditions are applied at walls and strongly influence near-wall gradients and acoustic streaming. In cryocooler geometries, walls are commonly modelled as either adiabatic or isothermal. An adiabatic wall condition implies zero normal heat flux:

$$\left.\frac{\partial T}{\partial n}\right|_{\text{wall}} = 0, \tag{7.10}$$

which is suitable for vacuum-jacketed sections or insulating regions. Conversely, an isothermal wall prescribes a fixed temperature:

$$T|_{\text{wall}} = T_{\text{wall}}. \tag{7.11}$$

In regions where heat exchange occurs, such as at cold or warm heat exchangers, a Robin-type condition may be used to model convective heat transfer:

$$-k\frac{\partial T}{\partial n} = h(T - T_{\infty}), \tag{7.12}$$

where h is the convective heat transfer coefficient and T_{∞} is the ambient or sink temperature (Incropera *et al* 2007).

Symmetry and geometric simplifications are employed to reduce computational cost while preserving physical behaviour. For instance, linear cryocooler geometries may use symmetry planes along the centerline, enforcing

$$\mathbf{u} \cdot \mathbf{n}|_{\text{sym}} = 0, \quad \left.\frac{\partial \phi}{\partial n}\right|_{\text{sym}} = 0, \tag{7.13}$$

for any scalar field ϕ. In axisymmetric components, such as regenerators and pulse tubes, the governing equations can be formulated in cylindrical coordinates to retain radial and axial structure while minimising dimensionality (Ferziger and Perić 2002, Swift 2002).

At the outlet, preventing acoustic wave reflections is vital to maintaining physical realism. This is particularly important in simulations spanning many acoustic cycles, where small reflection artifacts can accumulate. Several approaches exist for implementing non-reflecting outlet conditions. One widely used technique is to apply a damping region or sponge zone near the outlet, in which the momentum and energy equations are augmented with spatially varying attenuation terms. For example:

$$\frac{\partial(\rho\mathbf{u})}{\partial t} + \nabla \cdot (\rho\mathbf{u} \otimes \mathbf{u}) = -\nabla p + \nabla \cdot \boldsymbol{\tau} + \rho\mathbf{g} - \sigma(x)\rho\mathbf{u}, \tag{7.14}$$

where $\sigma(x)$ increases smoothly toward the boundary to dissipate outgoing waves (Tam and Dong 1996). Alternatively, methods based on characteristic wave theory or the use of Riemann invariants may be applied to match the acoustic impedance at the domain boundary and allow wave transmission without reflection (Poinsot and Lele 1992).

In summary, careful boundary specification, including oscillatory driving conditions, accurate thermal wall modelling, symmetry exploitation, and wave-transparent outlet designs, is essential for capturing key phenomena in oscillatory cryocooler flows, such as thermoacoustic wave propagation, acoustic streaming, and regenerator heat transfer. Subsequent case studies will illustrate these techniques in the simulation of Stirling and pulse tube cryocooler architectures.

7.3.3 Turbulence and transition modelling

Turbulence plays a complex and often critical role in the internal flow dynamics of cryocooler components, particularly in regions such as inertance tubes, pulse tubes, and high-Reynolds number regenerators. Accurately modelling the transition between laminar and turbulent regimes is essential for capturing the true nature of heat transfer, pressure loss, and acoustic wave distortion. The flow regime is typically classified using the Reynolds number, a dimensionless quantity defined as

$$\mathrm{Re}=\frac{\rho u_{\max} D}{\mu}, \tag{7.15}$$

where ρ is the fluid density, $u_{\max}$ is the maximum velocity amplitude, D is the characteristic hydraulic diameter, and μ is the dynamic viscosity of the working gas (usually helium). In oscillating flow systems, care must be taken when interpreting Reynolds number since the flow may exhibit both forward and reverse shear, leading to localised unsteady separation and reattachment.

A range of turbulence modelling strategies are employed in CFD simulations depending on the desired trade-off between fidelity and computational cost. The most widely used class of models is Reynolds-averaged Navier–Stokes (RANS), which introduces statistical averaging of the velocity field to close the Reynolds stress terms. Common two-equation RANS models include the k–ε model, which solves for the turbulent kinetic energy k and its dissipation rate ε, and the k–ω shear stress transport (SST) model, which blends near-wall and free-stream formulations for better prediction of boundary layer separation. These models are widely adopted in engineering-scale cryocooler simulations due to their robustness and computational efficiency (Wilcox 2006).

For flows in which large-scale coherent vortices dominate and transitional phenomena are critical, such as acoustic streaming or vortex shedding in inertance tubes, large eddy simulation (LES) provides a more accurate alternative. LES resolves the large-scale turbulent structures directly while modelling only the smaller subgrid-scale eddies. This allows better representation of transient dynamics at the expense of significantly higher computational cost. LES has been employed in high-fidelity research simulations to study energy cascade and acoustic-thermal coupling in pulse tube geometries (Kobayashi *et al* 2011).

In the limit of maximum fidelity, direct numerical simulation (DNS) solves the full, unfiltered Navier–Stokes equations without any turbulence modelling. DNS resolves all spatial and temporal scales of the flow, including the Kolmogorov microscale, and thus requires extremely fine meshes and small time steps. As a result, DNS is only feasible for simplified geometries and relatively low Reynolds numbers. Nevertheless, it serves as a vital tool for model validation and for building physical intuition about transition thresholds in cryogenic oscillatory flows (Moin and Mahesh 1998).

Due to the unique nature of oscillating flows in cryocoolers, where transition to turbulence may be intermittent and spatially localised, hybrid methods that blend laminar and turbulent models are increasingly used. These include detached eddy simulation (DES) and transitional RANS models, which incorporate empirical correlations or intermittency factors to predict laminar–turbulent breakdown (Langtry and Menter 2009). Such approaches are particularly valuable for predicting flow resistance in regenerators and performance degradation due to turbulent mixing.

In all cases, careful validation against experimental data or high-fidelity simulations is crucial, particularly because most turbulence models are calibrated for steady or unidirectional flows, rather than the oscillatory, compressible environment characteristic of cryogenic refrigeration systems.

7.3.4 Volume-averaged momentum equation

In porous media such as regenerators, the microscale flow through the packed matrix of screens or spheres is highly complex and computationally expensive to resolve directly. To facilitate tractable modelling while preserving the essential physics, a volume-averaged approach is employed, wherein the governing momentum equation is spatially averaged over a representative elementary volume (REV). This results in a continuum-scale description of the macroscopic flow behaviour within the porous domain.

Under the assumptions of steady-state, unidirectional flow and negligible compressibility effects, the volume-averaged form of the momentum equation simplifies to a Darcy–Forchheimer model, which incorporates both viscous and inertial resistance terms:

$$\frac{\partial p}{\partial x} = -\frac{\mu}{K}u - \rho\beta u^2, \tag{7.16}$$

where $\frac{\partial p}{\partial x}$ is the pressure gradient along the flow direction, μ is the dynamic viscosity of the fluid, ρ is the density, and u is the superficial velocity averaged over the porous volume. The parameter K is the permeability of the matrix, representing the resistance to viscous flow, and β is the Forchheimer coefficient, which accounts for inertial effects that become significant at moderate to high Reynolds numbers.

The first term on the right-hand side represents the classical Darcy resistance due to viscous drag in laminar flow through pores, while the second term captures non-linear losses due to flow acceleration, jetting, and turbulence-like phenomena at the pore scale. This model is widely used for regenerators in Stirling and pulse tube cryocoolers, where the Reynolds number typically lies in the range $\mathrm{Re} \sim 10^2$ to 10^3, rendering both viscous and inertial contributions important (Gedeon 1995, Radebaugh 2000).

Permeability K and Forchheimer coefficient β are geometry-dependent and are often determined empirically or from correlations based on the structure of the matrix. For example, in packed beds of spheres or sintered wire mesh, the widely used Ergun equation provides empirical expressions for both terms as functions of porosity ε, particle diameter d_p, and flow properties (Ergun 1952):

$$\frac{\partial p}{\partial x} = -\left[\frac{150\mu(1-\varepsilon)^2}{\varepsilon^3 d_p^2}\right]u - \left[\frac{1.75\rho(1-\varepsilon)}{\varepsilon^3 d_p}\right]u^2. \tag{7.17}$$

Comparison with the general Darcy–Forchheimer form allows direct identification of K and β in terms of pore-scale geometric parameters. This formulation is especially useful for time-domain or frequency-domain simulations that do not resolve the internal structure of the regenerator, but must accurately predict pressure drop and flow impedance.

The volume-averaged momentum equation provides an essential link between detailed microscale physics and computationally efficient macroscopic models of cryocooler components. It is routinely used in both Sage simulations and finite volume solvers (chapter 5), as well as in analytical treatments of thermoacoustic and harmonic models (chapters 4 and 6), where the regenerator is treated as a distributed resistance and reactance element.

Figure 7.3 illustrates the variation in pressure gradient magnitude across a porous medium, such as a regenerator matrix, as a function of superficial gas velocity.

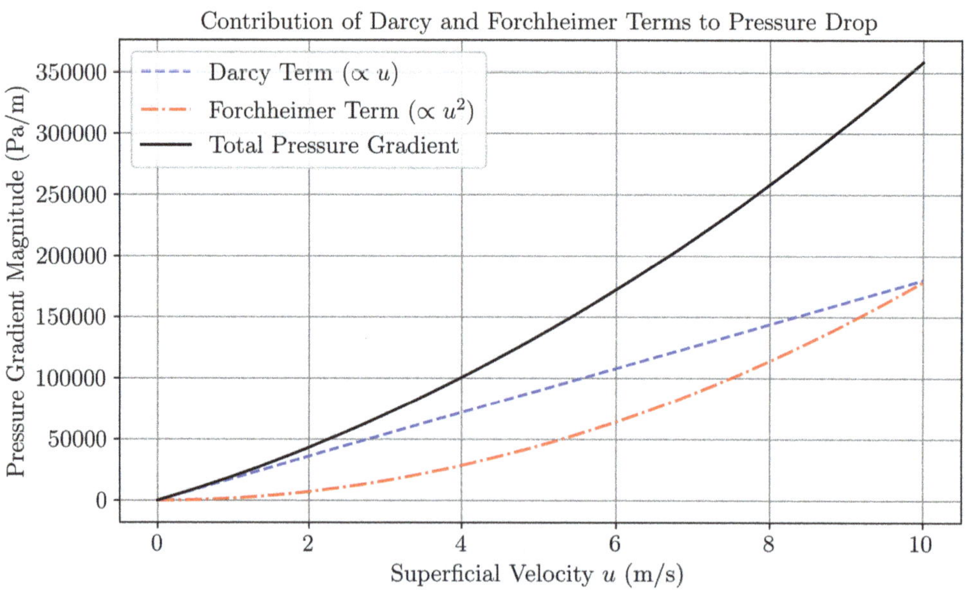

Figure 7.3. Pressure gradient magnitude as a function of superficial velocity u, illustrating the separate and combined contributions of the linear Darcy term and the quadratic Forchheimer term in a porous medium.

At small values of u, the viscous term dominates and the pressure gradient increases linearly, corresponding to Darcy's law. However, as the flow velocity increases, the quadratic Forchheimer term becomes increasingly significant, capturing the inertial effects that arise from non-laminar streamline deviations and flow acceleration through narrow pores. This transition between flow regimes is important for cryocooler components such as regenerators, where oscillatory flow spans a range of Reynolds numbers over each cycle. The combined Darcy–Forchheimer formulation is widely used in CFD and system-level models to characterize flow impedance in porous media subjected to oscillatory flow (Gedeon 1995).

7.3.5 LTNE heat transfer in porous media

In porous media subjected to oscillatory or transient thermal loading, such as regenerators in cryocoolers, the assumption of local thermal equilibrium (LTE) between the solid matrix and the fluid phase can break down. This is particularly relevant when the thermal diffusivities of the solid and fluid differ significantly, or when rapid temporal variations prevent sufficient interphase heat exchange. To resolve this, the local thermal non-equilibrium (LTNE) model treats the solid and fluid phases as possessing independent temperatures, T_s and T_f, respectively.

The energy balance for the solid phase is governed by

$$\rho_s c_s \frac{\partial T_s}{\partial t} = h_{sf}(T_f - T_s) + k_s \nabla^2 T_s, \tag{7.18}$$

where ρ_s and c_s denote the density and specific heat capacity of the solid phase, k_s is the effective thermal conductivity of the solid skeleton, and h_{sf} is the volumetric interfacial heat transfer coefficient characterising energy exchange between the fluid and solid domains.

The corresponding energy equation for the fluid phase accounts for convective transport due to bulk flow and is given by

$$\rho_f c_f \frac{\partial T_f}{\partial t} + \rho_f c_f \mathbf{u} \cdot \nabla T_f = h_{sf}(T_s - T_f) + k_f \nabla^2 T_f, \tag{7.19}$$

where ρ_f, c_f, and k_f are the density, specific heat capacity, and effective thermal conductivity of the fluid, respectively, and $\mathbf{u}$ is the local fluid velocity vector. The term $h_{sf}(T_s - T_f)$ enforces the energy exchange based on the local temperature difference between the two phases.

These coupled partial differential equations describe how the fluid and solid temperatures evolve differently in time and space, allowing for more accurate modelling of regenerator dynamics where steep thermal gradients exist during the oscillatory flow cycle. The LTNE framework has been widely employed in the analysis of packed beds, regenerators, and metal foams (Nield and Bejan 2006). It provides a crucial extension over the simpler LTE assumption, in particular in high-frequency regimes or when evaluating transient startup behaviour in cryogenic systems.

7.4 Practical applications and case studies

7.4.1 Example 1: Pulse tube with inertance tube

To illustrate the detailed thermo-fluid dynamics that emerge in oscillating cryogenic systems, we consider a time-domain CFD simulation of a pulse tube cryocooler equipped with an inertance tube. This example showcases both the spatial and temporal evolution of key flow quantities, namely pressure, velocity, and entropy, and provides insight into the mechanisms of energy transport, flow reversal, and boundary layer formation. Figure 7.4 presents axial velocity fields within the pulse tube at two characteristic time points during the oscillation cycle: peak compression and peak expansion. These snapshots are extracted from a high-fidelity transient simulation, where the governing equations for mass, momentum, and energy conservation are solved under compressible, unsteady flow conditions with no-slip adiabatic walls. At peak compression, the gas is driven toward the cold end, resulting in strong boundary layer acceleration and axial bulk flow in the positive direction. Conversely, at peak expansion, the flow direction reverses and the boundary layers detach, initiating vortex formation and flow separation near the tube ends. These dynamics are essential in understanding entropy generation and acoustic-to-thermal energy conversion within the pulse tube (Radebaugh 2000).

To quantitatively characterize the flow behaviour along the tube axis, figure 7.5 plots the amplitude of pressure and velocity oscillations as a function of axial position over one full cycle. The pressure amplitude decreases from the hot end to the cold end, consistent with the impedance imposed by the regenerator and inertance tube. In contrast, the velocity amplitude peaks near the midpoint of the pulse tube, where the standing wave profile achieves its maximum displacement. These profiles are extracted using a steady-periodic averaging technique, whereby flow variables are decomposed into their fundamental Fourier components once the system reaches limit-cycle behaviour. Mathematically, the axial velocity may be represented as

$$u(x, t) = \mathrm{Re}\{\tilde{u}(x)e^{i\omega t}\}, \tag{7.20}$$

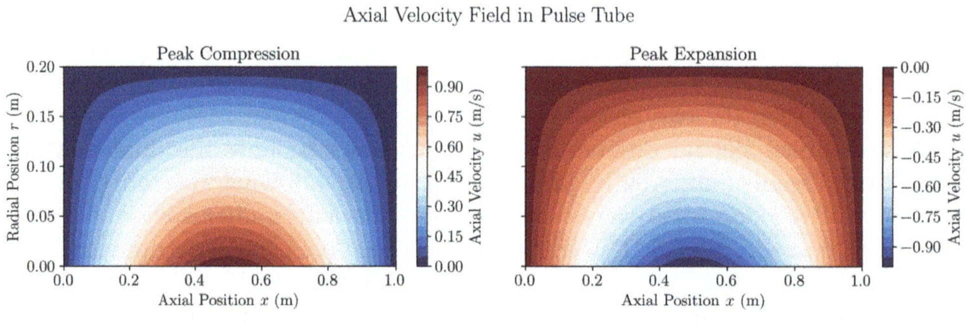

Figure 7.4. Axial velocity field in a pulse tube at two instants during a full oscillation cycle. Left: peak compression phase, where gas flows toward the cold end. Right: peak expansion phase, where flow reverses direction. The formation of boundary layers and the parabolic core flow profile are evident.

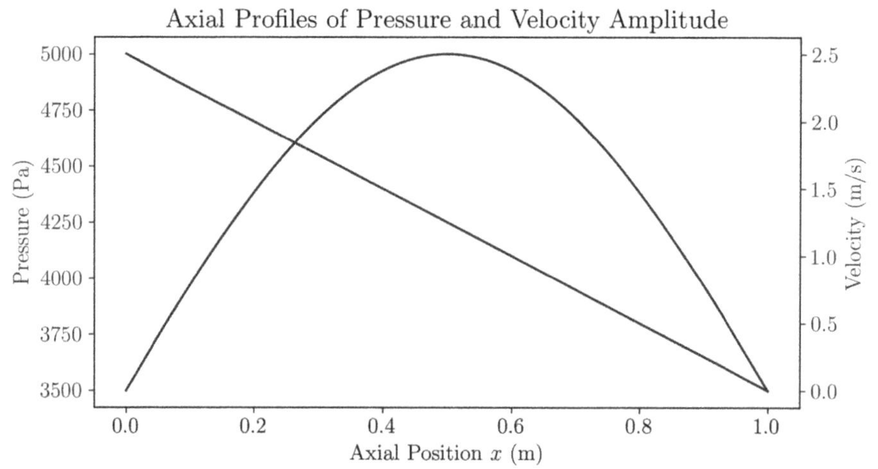

Figure 7.5. Axial profiles of pressure and velocity amplitude within the pulse tube over a representative oscillation cycle. The pressure amplitude (blue, left axis) decreases monotonically due to wave attenuation or impedance matching at the ends. The velocity amplitude (red dashed, right axis) follows a parabolic profile, reaching its maximum at the tube midpoint, as is characteristic of standing wave behaviour in oscillatory thermofluid systems.

where $\tilde{u}(x)$ is the complex amplitude determined from simulation data, and ω is the angular frequency of oscillation. A similar representation is used for the pressure and entropy fields.

Of particular interest is the local entropy generation due to thermal and viscous dissipation, which can be computed using the expression

$$\dot{s}_{\text{gen}} = \frac{k}{T^2}(\nabla T)^2 + \frac{\Phi}{T}, \tag{7.21}$$

where k is the thermal conductivity, T is the local temperature, and Φ is the viscous dissipation function, given by

$$\Phi = 2\left(\frac{\partial u}{\partial x}\right)^2 + \left(\frac{\partial v}{\partial y}\right)^2 + \left(\frac{\partial w}{\partial z}\right)^2 + \cdots, \tag{7.22}$$

including all components of the strain-rate tensor for three-dimensional flow. These terms dominate near the walls due to steep velocity gradients and in regions of flow reversal, particularly close to the open end of the pulse tube and near the inertance junction.

In summary, this example illustrates the rich interplay between oscillatory flow dynamics, boundary layer evolution, and entropy generation in cryogenic pulse tube systems. The spatial and temporal information obtained from time-resolved CFD simulations provides critical insight into the performance limitations and optimisation opportunities in regenerative cryocooler design. Detailed visualisations such as those shown in figures 7.4 and 7.5 also serve as valuable benchmarks for validating simplified phasor or one-dimensional models discussed in earlier chapters.

7.4.2 Example 2: Regenerator flow with Ergun equation

The flow of gas through a cryogenic regenerator matrix can be described as transport through a porous medium exhibiting both viscous and inertial resistance. A widely used empirical correlation for modelling this behaviour is the Ergun equation, which combines Darcy-like linear drag and Forchheimer-like quadratic drag into a unified expression for steady-state, incompressible flow. The equation is given by

$$\Delta p = \frac{150\mu(1-\varepsilon)^2}{\varepsilon^3 d_\mathrm{p}^2} L u + \frac{1.75\rho(1-\varepsilon)}{\varepsilon^3 d_\mathrm{p}} L u^2, \tag{7.23}$$

where μ is the dynamic viscosity, ρ is the gas density, ε is the porosity of the matrix, d_p is the effective pore or particle diameter, L is the regenerator length, and u is the superficial velocity. The first term corresponds to viscous (Darcy) losses, while the second accounts for inertial (Forchheimer) contributions arising from streamline curvature and local flow acceleration (Ergun 1952).

Figure 7.6 compares the Fanning friction factor predicted by the Ergun equation against synthetic data obtained from CFD simulations of oscillatory flow through a regenerator mesh. The results are presented in nondimensional form, with the horizontal axis showing the pore Reynolds number

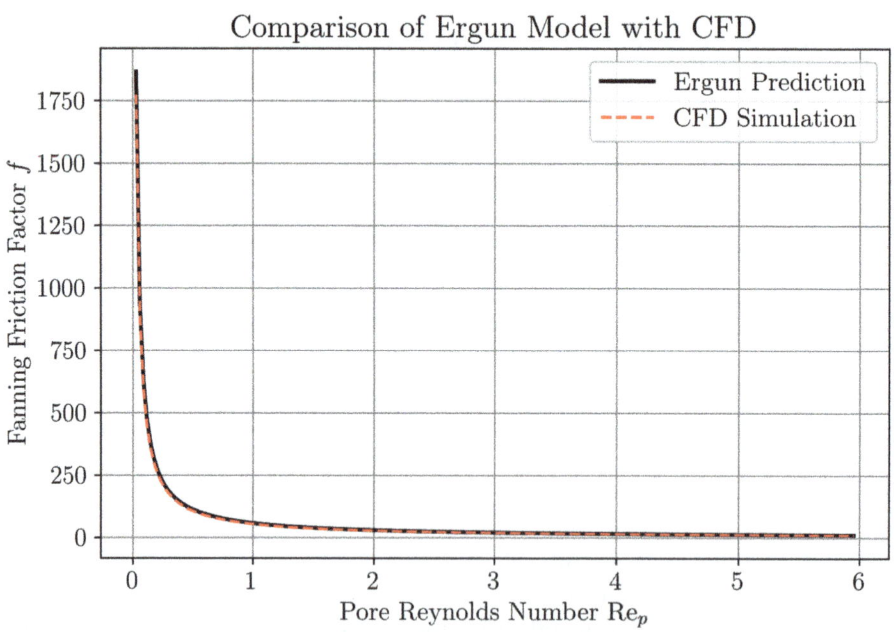

Figure 7.6. Comparison between Ergun equation prediction and CFD results for flow through a regenerator mesh. Data are plotted as nondimensional Fanning friction factor versus Reynolds number. CFD results show good agreement in the transitional regime but deviate at high Re_p due to oscillatory and compressible effects.

$$\mathrm{Re_p} = \frac{\rho u d_\mathrm{p}}{\mu}, \tag{7.24}$$

and the vertical axis representing the Fanning friction factor

$$f = \frac{\Delta p}{2L} \cdot \frac{d_\mathrm{p}}{\rho u^2}. \tag{7.25}$$

The Ergun model agrees well with simulation data in the transitional regime, but discrepancies arise at higher $\mathrm{Re_p}$ due to unsteady effects and compressibility not accounted for in the original correlation.

Beyond momentum loss, accurate thermal modelling within regenerators is critical for predicting cryocooler performance. Traditional models often assume LTE, where the fluid and solid matrix share a common temperature field. However, this assumption breaks down when transient or spatially non-uniform heat exchange occurs, in particular in high-frequency oscillatory flow where limited thermal penetration prevents full equilibration.

To capture such effects, the LTNE model treats the gas and solid phases separately, each with its own energy equation coupled through an interfacial heat transfer coefficient. Figure 7.7 shows a comparison of temperature fields computed under LTE and LTNE conditions. The LTE model yields a single smooth temperature field, whereas the LTNE simulation reveals significant thermal gradients between the gas and matrix, particularly near boundaries and at flow reversal regions. This distinction becomes increasingly important in regenerators operating at high frequencies, where phase-lagged thermal responses can degrade performance (Nield and Bejan 2006).

Together, these results emphasize the necessity of incorporating both accurate flow resistance models and appropriate thermal closure assumptions in regenerator simulations. The Ergun equation provides a reliable baseline for modelling pressure drop in porous media, while LTNE-based heat transfer formulations are essential for resolving inter-phase temperature disparities under oscillatory flow conditions.

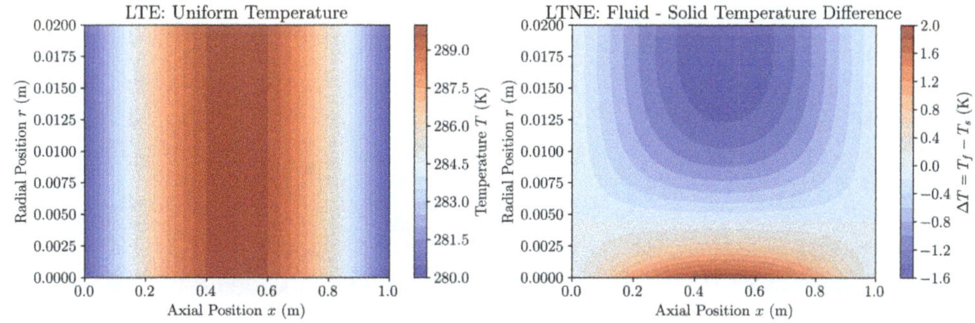

Figure 7.7. Temperature field contours across a regenerator cross-section at steady-periodic state. Left: LTE model with shared fluid-solid temperature. Right: LTNE model showing distinct temperature fields and lag between gas and solid phases.

7.4.3 Example 3: Onset of turbulence

The transition from laminar to turbulent flow in cryocooler components such as the pulse tube can significantly alter both pressure drop and heat transfer characteristics. While the classical Reynolds number threshold for turbulence in pipe flow is around 2300, oscillatory flows exhibit more complex behaviour, with critical Reynolds numbers that depend on geometry, frequency, and waveform shape.

To visualize the onset of turbulence, we perform unsteady CFD simulations of a representative pulse tube geometry across a range of operating conditions. Figure 7.8 shows vorticity magnitude contours at the time of peak inertial acceleration, revealing the development of coherent vortex structures near the walls and junctions. These flow features are indicative of Kelvin–Helmholtz instabilities, which trigger breakdown into smaller turbulent eddies as the oscillation amplitude increases.

Complementary to vorticity, the Q-criterion can also be employed to identify regions where rotation dominates strain rate, providing a more selective indicator of vortex cores. High-Q regions correspond to zones of concentrated rotational energy and are typically found near geometric transitions and boundary layers. These diagnostic fields are critical for assessing the validity of turbulence models applied to pulse tube simulations.

To quantify the transition, we compute the turbulent kinetic energy (TKE) distribution along the pulse tube length for both laminar and k–ε turbulence models. The TKE is defined as

$$\text{TKE}=\frac{1}{2}(\overline{u'^2}+\overline{v'^2}+\overline{w'^2}), \tag{7.26}$$

where the overbar denotes time-averaging over one acoustic period and the primes indicate fluctuating velocity components. Figure 7.9 presents the normalised axial distribution of TKE, showing that the laminar model predicts negligible kinetic energy dissipation, while the turbulent model captures a sharp rise in TKE

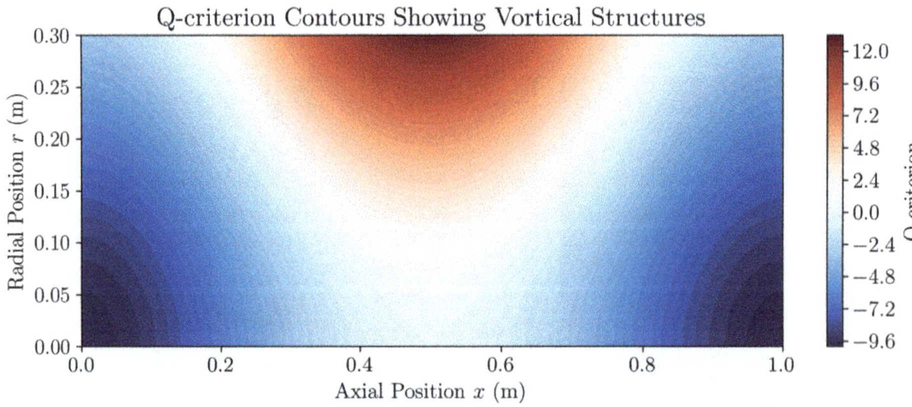

Figure 7.8. Vorticity magnitude and Q-criterion contours at peak inertial phase, visualising the onset of turbulence and vortex core development in a pulse tube geometry.

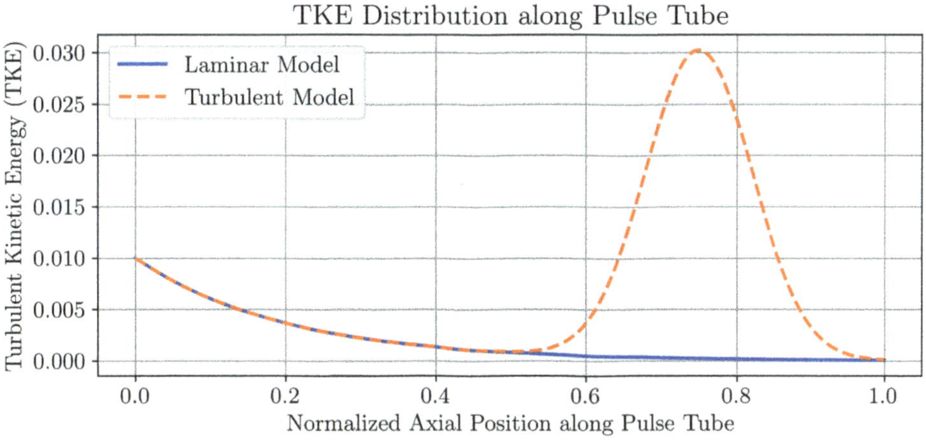

Figure 7.9. TKE profile along the pulse tube axis comparing laminar and turbulent k–ε models. TKE rises sharply near the junction region as flow instability sets in.

downstream of the junction region. The peak occurs near $x/D \approx 0.75$, consistent with observed vortex shedding in the unsteady simulation.

These results emphasize the importance of incorporating appropriate turbulence models in high-fidelity cryocooler simulations, particularly under conditions of high acoustic Reynolds number and steep velocity gradients. While full DNS remains impractical for most engineering applications, RANS models such as k–ε offer a computationally efficient means of capturing the average effects of turbulence in oscillating cryogenic flows.

7.5 Coupling with heat transfer and thermodynamics

The performance of pulse tube cryocoolers can be more comprehensively evaluated by coupling the momentum and energy equations to quantify local and global thermodynamic irreversibilities. Specifically, entropy generation provides a rigorous metric for evaluating the second-law efficiency of cryogenic systems. The volumetric rate of entropy generation, $\dot{S}_{\text{gen}}$, can be expressed as the sum of viscous and thermal contributions:

$$\dot{S}_{\text{gen}} = \int_V \left[\frac{\Phi}{T} + \frac{k}{T^2} (\nabla T)^2 \right] \mathrm{d}V, \tag{7.27}$$

where Φ is the viscous dissipation function (proportional to the square of the velocity gradient tensor), k is the thermal conductivity, T is the local temperature, and $(\nabla T)^2$ represents the magnitude of the temperature gradient squared. This expression captures the irreversible losses due to both internal friction and heat conduction down a temperature gradient.

Figure 7.10 shows a spatial heatmap of entropy generation within the pulse tube region, computed from CFD results of the coupled flow and thermal fields. The plot highlights zones of high entropy production, particularly near the tube ends where

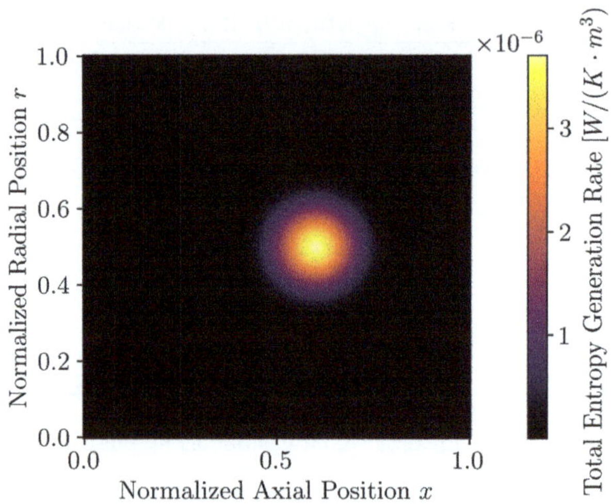

Figure 7.10. Entropy generation rate map in the pulse tube showing spatial distribution of local irreversibilities. Zones near the boundaries experience elevated entropy production due to strong velocity gradients and thermal conduction.

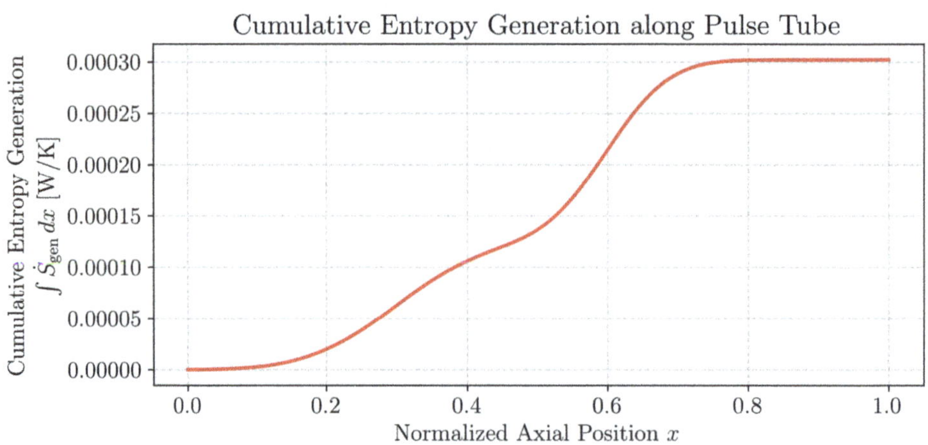

Figure 7.11. Cumulative entropy generation as a function of normalised axial position from the hot to cold end. The majority of irreversibilities originate near the thermal reservoirs and boundary layers.

strong temperature gradients and oscillatory shear drive irreversibility. These zones correlate with the locations of acoustic streaming and regenerator proximity, emphasising the need for targeted thermal management.

To further illustrate the thermodynamic cost of cryocooler operation, figure 7.11 plots the cumulative entropy generation as a function of axial position from hot to cold end. The distribution reveals that the majority of entropy is produced near the boundaries, consistent with regions of large velocity divergence and thermal diffusion. These diagnostics support the use of entropy generation minimisation as an objective function in cryocooler optimisation strategies.

7.6 Software and solver considerations

Numerical simulation of cryogenic oscillatory systems relies heavily on the choice of software platform and solver configuration. Several commercial and open-source CFD packages are commonly employed depending on the modelling needs, boundary conditions, and physical coupling complexity. For structured geometries and high-fidelity turbulence modelling, ANSYS Fluent offers a robust framework with mature implementations of RANS, LES, and transition models. Its pressure-based solvers are well-suited for resolving compressible and low-Mach flows typical of pulse tube cryocoolers, and built-in functionality for oscillating boundary conditions allows direct transient modelling of acoustic fields (ANSYS 2024).

Alternatively, OpenFOAM provides a fully open-source environment where custom equations can be implemented directly into the solver structure. This flexibility is especially valuable in modelling porous media, regenerators, and user-defined heat transfer coefficients, and it enables the inclusion of source terms such as viscous heating or entropy generation based on specific thermodynamic formulations. OpenFOAM's PIMPLE algorithm (a hybrid of PISO and SIMPLE) is particularly effective for unsteady simulations with strong pressure–velocity coupling (Weller *et al* 1998).

For problems requiring multiphysics coupling—such as heat conduction in solids, conjugate heat transfer between fluids and walls, or electromagnetic-thermal coupling in active elements, COMSOL Multiphysics is a suitable option. Its finite element formulation allows highly flexible domain definitions and accurate resolution of temperature fields and entropy transport within complex boundary conditions (COMSOL 2023).

Beyond software selection, the configuration of solver strategy plays a crucial role in both computational cost and physical accuracy. For oscillatory cryogenic systems, two primary approaches are commonly adopted. The first is direct time-domain integration over many cycles using transient solvers until periodic steady-state is achieved. This method is suitable for capturing transient startup behaviour, nonlinear harmonics, and phase-resolved interactions in strongly coupled systems. The second approach is the use of steady-periodic solvers, which assume purely periodic boundary conditions and compute only a single representative cycle, significantly reducing simulation time in systems where transients have died out (Radebaugh 2000).

Additionally, one must consider whether to solve the fluid and energy equations in a *segregated* or *coupled* fashion. Segregated solvers handle pressure, momentum, and energy equations sequentially, iterating between them until convergence. While computationally cheaper per iteration, this approach may struggle with stiff coupling between fields in high-impedance or low-porosity regions. Coupled solvers, on the other hand, solve all equations simultaneously using block-matrix approaches or implicit schemes. Although more demanding per iteration, they often yield faster overall convergence in strongly interdependent domains.

To ensure numerical accuracy, all solvers must be tested for mesh independence. The influence of grid resolution on key thermodynamic outputs, such as cold-tip temperature, pressure drop, and entropy generation rate, is assessed through grid sensitivity studies. These tests, discussed in chapter 5, confirm whether numerical diffusion or under-resolution affects the reliability of physical predictions. Ideally, simulations should converge toward an asymptotic solution as mesh is refined, with error norms decreasing at a rate consistent with the underlying discretisation order.

7.7 Summary

Hydrodynamic modelling techniques presented in this chapter go significantly beyond the one-dimensional and phasor-based approximations introduced earlier in the book. These extended formulations enable spatial and temporal resolution of complex field phenomena such as boundary layer development, vortex dynamics, and acoustic streaming. Computational fluid dynamics (CFD), in particular, provides a powerful framework for simulating flow separation, local instabilities, and turbulent transitions that influence performance and efficiency in cryogenic pulse tube and Stirling systems. By resolving transient inertial effects and oscillatory phase interactions, CFD-based tools offer insight into loss mechanisms that are inaccessible through simplified models.

Within the regenerator, the incorporation of porous media models, including Ergun-based formulations for flow resistance and local thermal non-equilibrium (LTNE) energy balances, enhances the fidelity of predicted temperature and pressure fields. These models better capture the dynamic thermal interactions between the solid matrix and the gas phase, especially under high-frequency or large-amplitude conditions where equilibrium assumptions fail.

Crucially, the coupling of momentum and energy equations also permits direct evaluation of entropy generation throughout the system. The integration of viscous dissipation and thermal conduction terms enables local and global assessments of thermodynamic irreversibility, providing a quantitative basis for optimising cryocooler geometry and operating conditions. By identifying zones of high entropy production, designers can implement targeted modifications to reduce thermal losses, improve phase alignment, and enhance overall system efficiency.

Taken together, these hydrodynamic and thermodynamic modelling capabilities provide a rigorous foundation for advanced cryocooler design and performance prediction.

References

ANSYS 2024 *ANSYS Fluent Theory Guide* (ANSYS Inc.)

Backhaus S and Swift G W A 2000 Thermoacoustic-Stirling heat engine: detailed study *J. Acoust. Soc. Am.* **107** 3148–66

COMSOL 2023 *COMSOL Multiphysics User's Guide* (COMSOL AB)

Ergun S 1952 Fluid flow through packed columns *Chem. Eng. Prog.* **48** 89–94

Ferziger J H and Perić M 2002 *Computational Methods for Fluid Dynamics* (Berlin: Springer)

Gedeon D 1995 DC gas flows in Stirling and pulse tube cryocoolers *Cryocoolers* **vol 9** (New York: Springer) pp 385–92

Gedeon D 1995 Distributed loss model for Stirling and pulse tube cryocoolers *Cryocoolers* **vol 9** (New York: Springer) pp 385–92

Gedeon D 1998 Component physics in pulse tube refrigerators *Cryocoolers* **vol 10** (New York: Springer) pp 187–97

Incropera F P, DeWitt D P, Bergman T L and Lavine A S 2007 *Fundamentals of Heat and Mass Transfer* 6th edn (New York: Wiley)

Issa R I 1986 Solution of the implicitly discretised fluid flow equations by operator splitting *J. Comput. Phys.* **62** 40–65

Kobayashi H, Fukuda K and Matsubara K 2011 Large eddy simulation of thermoacoustic oscillations in a pulse tube cryocooler *Cryogenics* **51** 364–72

Langtry R B and Menter F R 2009 Correlation-based transition modelling for unstructured parallelised computational fluid dynamics codes *AIAA J.* **47** 2894–906

Moin P and Mahesh K 1998 Direct numerical simulation: a tool in turbulence research *Annu. Rev. Fluid Mech.* **30** 539–78

Nield D A and Bejan A 2006 *Convection in Porous Media* 3rd edn (New York: Springer)

Patankar S V 1980 *Numerical Heat Transfer and Fluid Flow* (London: Hemisphere Publishing)

Poinsot T J and Lelef S K 1992 Boundary conditions for direct simulations of compressible viscous flows *J. Comput. Phys.* **101** 104–29

Radebaugh R 2000 Development of the pulse tube refrigerator as an efficient and reliable cryocooler *Proc. Inst. Refrig.* **96** 11–29

Rawlings T 2022 *Numerical Modelling of Stirling Cryocoolers* (London: University College London)

Swift G W 1988 Thermoacoustics: a unifying perspective for some engines and refrigerators *J. Acoust. Soc. Am.* **84** 1145–80

Tam C K W and Dong Z 1996 Radiation and outflow boundary conditions for direct computation of acoustic and flow disturbances in a nonuniform mean flow *J. Comput. Acoust.* **4** 175–201

Van Sciver S W 2012 *Helium Cryogenics* 2nd edn. (New York: Springer)

Versteeg H K and Malalasekera W 2007 *An Introduction to Computational Fluid Dynamics: The Finite Volume Method* 2nd edn (Harlow: Pearson Education)

Weller H G, Tabor G, Jasak H and Fureby C 1998 A tensorial approach to computational continuum mechanics using object-oriented techniques *Comput. Phys.* **12** 620–31

Wilcox D C 2006 *Turbulence Modelling for CFD* 3rd edn (La Cañada, CA: DCW Industries)

Chapter 8

Data analysis and optimisation methods

8.1 Signal processing in the lab

8.1.1 Time-domain and frequency-domain representations

Cryocooler systems generate oscillatory signals such as pressure, mass flow rate, voltage, or displacement, all of which carry essential information about system performance. These signals are inherently time-varying and often periodic, particularly when driven by sinusoidal compressor motions or when characterised in steady-periodic operating regimes. It is therefore vital to understand both the time-domain and frequency-domain representations of such signals.

In the time domain, a signal $x(t)$ is expressed as a function of time, capturing instantaneous amplitudes. This is the native form in which data are acquired in the laboratory. However, time-domain plots can obscure important spectral features, such as harmonics, phase shifts, and noise signatures, which become more accessible through frequency-domain analysis.

For a periodic function $x(t)$ with period $T = \frac{2\pi}{\omega}$, the Fourier series representation decomposes the signal into its constituent harmonics:

$$x(t) = a_0 + \sum_{n=1}^{\infty}[a_n \cos(n\omega t) + b_n \sin(n\omega t)],$$

where the Fourier coefficients are given by

$$a_0 = \frac{1}{T}\int_0^T x(t)\,\mathrm{d}t, \quad a_n = \frac{2}{T}\int_0^T x(t)\cos(n\omega t)\,\mathrm{d}t, \quad b_n = \frac{2}{T}\int_0^T x(t)\sin(n\omega t)\,\mathrm{d}t.$$

This decomposition allows each harmonic to be interpreted as an orthogonal component in function space, enabling spectral isolation of dominant physical processes such as the fundamental compressor frequency and higher-order nonlinear effects.

Parseval's theorem provides a useful identity relating the time-domain energy of the signal to the sum of its squared frequency components:

doi:10.1088/978-0-7503-4826-3ch8

$$\langle x^2(t)\rangle = \frac{1}{T}\int_0^T x^2(t)\,\mathrm{d}t = a_0^2 + \frac{1}{2}\sum\nolimits_{n=1}^{\infty}\left(a_n^2 + b_n^2\right),$$

where the left-hand side represents the mean-square value of the signal over a period. This is particularly useful in cryocooler performance evaluation, where RMS values of pressure and velocity contribute directly to acoustic power estimates.

In practical applications, especially with digitised signals, the Fourier transform is computed numerically using the fast Fourier transform (FFT), which enables efficient conversion of time-domain data to a discrete spectrum of frequencies. The discrete Fourier transform X_k of a sampled signal x_n of length N is given by

$$X_k = \sum\nolimits_{n=0}^{N-1} x_n\, \mathrm{e}^{-j2\pi kn/N}, \quad k = 0, 1, \ldots, N-1.$$

This representation enables identification of spectral peaks, energy content, and phase relationships among oscillating cryocooler subsystems.

In cryogenic experimental set-ups, frequency-domain analysis is routinely used in lock-in amplification, noise diagnostics, and identifying nonlinear behaviours such as valve chatter or regenerator flow asymmetry. Moreover, frequency content informs optimal filtering strategies for signal conditioning prior to data logging or feedback control.

Ultimately, mastery of both time-domain and frequency-domain perspectives is essential for extracting physically meaningful metrics from experimental data, validating simulation models, and diagnosing system anomalies. Figure 8.1 illustrates a representative example of signal analysis in cryocooler diagnostics. The left panel shows a synthetic pressure waveform in the time domain, combining a 60 Hz

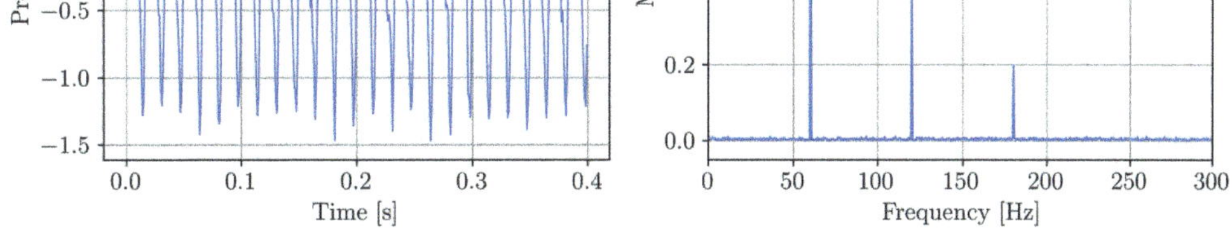

Figure 8.1. Time-domain (left) and frequency-domain (right) representations of a synthetic cryocooler pressure signal composed of a 60 Hz fundamental and its harmonics, with added Gaussian noise. The time trace exhibits a distorted sinusoid characteristic of non-sinusoidal periodic driving, while the FFT spectrum reveals strong peaks at 60 Hz, 120 Hz, and 180 Hz corresponding to harmonic content.

fundamental component with higher harmonics and additive noise. The resulting signal displays non-sinusoidal periodic behaviour, which could arise from valve dynamics or nonlinear compression effects. This complexity is difficult to interpret from the time trace alone.

The right panel shows the corresponding frequency-domain representation obtained via the FFT. The spectrum reveals sharp peaks at 60 Hz, 120 Hz, and 180 Hz, clearly identifying the harmonic content of the signal. These harmonics confirm the presence of waveform distortion and provide insight into mechanical or acoustic nonlinearity in the system. The frequency-domain plot also highlights the presence of broadband noise, visible as a flat floor across high frequencies, which may originate from sensor electronics or ambient environmental interference.

This example demonstrates the value of Fourier analysis in isolating dominant periodic components and diagnosing the source of waveform irregularities in cryocooler operation.

8.1.2 Fourier transform and spectral analysis

While Fourier series are well-suited to periodic signals, more general transient or non-periodic signals encountered in cryocooler diagnostics require the use of the Fourier transform. The continuous Fourier transform expresses a time-domain signal $x(t)$ as an integral of complex exponentials:

$$X(f) = \int_{-\infty}^{\infty} x(t)\, \mathrm{e}^{-j2\pi ft}\, \mathrm{d}t,$$

where $X(f)$ is a complex-valued function representing the amplitude and phase of the frequency component f. This transformation provides access to the full frequency spectrum, allowing identification of narrowband energy content, noise signatures, and spectral asymmetries.

The inverse Fourier transform reconstructs the time-domain signal from its spectral representation:

$$x(t) = \int_{-\infty}^{\infty} X(f)\, \mathrm{e}^{j2\pi ft}\, \mathrm{d}f.$$

In laboratory measurements, signals are sampled and finite in duration, so the discrete Fourier transform (DFT) is used, as given in section 8.1.1.

This discretisation yields a frequency-domain representation with resolution $\Delta f = \frac{1}{N\Delta t}$, limited by sampling rate and record length. Efficient computation is achieved using the FFT algorithm, which scales as $\mathcal{O}(N \log N)$ (Cooley and Tukey 1965).

Because the DFT assumes periodic extension of the sampled signal, discontinuities at the edges of the time window can introduce spectral leakage: artificial spreading of energy across nearby frequencies. This is particularly problematic for non-sinusoidal or transient waveforms, where abrupt endpoints bias the spectrum.

To mitigate this, window functions such as the Hann, Hamming, or Blackman windows are applied to taper the signal smoothly to zero at the edges:

$$x_n^{(\text{windowed})} = w_n \cdot x_n,$$

where $w_n \in [0, 1]$ is the window coefficient at sample n. These windows reduce side-lobe leakage at the cost of frequency resolution, a tradeoff known as the resolution–leakage dilemma.

Windowing is especially important in cryocooler performance testing when analysing the frequency response of pressure transducers, thermocouples, or displacement sensors under steady-state or modulated loading. Spectral analysis also supports identification of spurious components such as electromagnetic interference (EMI), resonance modes, or structural vibrations, which may compromise system performance or measurement integrity.

The FFT spectrum may be interpreted in terms of magnitude and phase:

$$|X(f)| = \sqrt{\mathrm{Re}(X(f))^2 + \mathrm{Im}(X(f))^2}, \quad \phi(f) = \tan^{-1}\left(\frac{\mathrm{Im}(X(f))}{\mathrm{Re}(X(f))}\right),$$

enabling quantification of gain and phase lag in response to harmonic excitation; parameters fundamental to characterising acoustic impedance and control system behaviour.

In many experiments, narrowband excitation or lock-in amplification is employed, and the FFT verifies that the signal is dominated by a single frequency component. Broadband spectra, on the other hand, provide insight into transients, valve dynamics, or coupling between mechanical and thermal modes.

8.1.3 Signal decomposition and filtering

In cryocooler experiments, measurement signals are frequently corrupted by noise, transients, or unwanted harmonics. Effective signal decomposition and filtering strategies are thus critical for isolating meaningful physical information, particularly when measuring small temperature oscillations, acoustic pressure waves, or displacer motions in the presence of environmental interference.

Signal decomposition refers to the separation of a signal into components according to frequency content. In the frequency domain, this is achieved by selectively retaining certain bands via digital or analog filtering. For example, low-pass filters preserve low-frequency components while attenuating high-frequency noise, whereas band-pass filters isolate a narrow range around a resonance or excitation frequency.

A common first-order low-pass filter has the transfer function:

$$H(f) = \frac{1}{1 + j\left(\frac{f}{f_c}\right)},$$

where f_c is the cutoff frequency, above which signal attenuation increases with frequency. This filter attenuates frequencies beyond f_c by a factor of $1/\sqrt{2}$ at the

-3 dB point. In the time domain, the filter corresponds to an exponential moving average and smooths out rapid variations, allowing clearer identification of quasi-steady-state trends.

The corresponding magnitude and phase response of the filter are

$$|H(f)| = \frac{1}{\sqrt{1 + \left(\frac{f}{f_c}\right)^2}}, \qquad \phi(f) = -\tan^{-1}\left(\frac{f}{f_c}\right),$$

showing that higher frequencies are both attenuated and phase-shifted. In multi-stage systems, cascaded filters may be applied to steepen roll-off characteristics or suppress specific harmonics more effectively.

Filtering is routinely employed to post-process lock-in amplifier outputs, where a narrow bandwidth around the excitation frequency is extracted to enhance signal-to-noise ratio. For instance, in thermal phase lag measurements using thermocouples, the fundamental frequency component is preserved while rejecting broadband background thermal fluctuations. Similarly, in laser displacement sensors monitoring piston or displacer motion, harmonic filtering isolates the mechanical stroke envelope from structural vibration modes or electrical switching noise.

Digital implementations of filters rely on convolution in the time domain or multiplication in the frequency domain. Finite impulse response (FIR) and infinite impulse response (IIR) designs are selected based on phase linearity, computational efficiency, and stability constraints. For example, zero-phase forward-reverse filtering is often applied in MATLAB or Python to avoid group delay in frequency-domain diagnostics.

Another approach is empirical mode decomposition (EMD), which decomposes a signal into intrinsic mode functions without relying on predefined basis functions. While computationally intensive, EMD can reveal subtle nonlinear or nonstationary features in cryocooler sensor signals, such as delayed thermal response or dynamic load coupling.

Ultimately, filtering and decomposition are essential tools in a cryogenicist's analytical arsenal. They enable accurate extraction of amplitude, phase, and spectral content from raw data, forming the basis for impedance calculations, efficiency metrics, and comparison with theoretical models.

8.2 Regression and curve fitting

Quantitative relationships between input and output variables in cryocooler systems are often inferred from experimental or simulated datasets. Regression and curve fitting techniques enable the construction of surrogate models, empirical trendlines, and system response approximations that can be embedded into control loops or design tools.

8.2.1 Linear and polynomial regression

In the simplest case, linear regression estimates the parameters β that best fit a model of the form:

$$\mathbf{y} = X\beta + \varepsilon,$$

where $\mathbf{y}$ is the vector of observed outputs, X is the design matrix containing input features (e.g. compressor frequency, input current), and ε is the error term. The least-squares estimator minimises the sum of squared residuals:

$$\min_{\beta} \left\| \mathbf{y} - X\beta \right\|^2.$$

The solution is obtained analytically by solving the normal equations:

$$\hat{\beta} = (X^T X)^{-1} X^T \mathbf{y}.$$

Polynomial regression extends this framework by including nonlinear basis functions of the inputs, such as x^2, x^3, or cross-terms, enabling capture of curved trends. In cryogenic systems, this is commonly used to fit second- or third-order relationships between cooling power and control parameters such as valve delay, regenerator fill factor, or input voltage.

8.2.2 Nonlinear curve fitting

Many relationships in cryocooler systems are inherently nonlinear, especially when involving thermodynamic, fluid, or acoustic behaviour. In these cases, nonlinear regression is applied to fit models of the form:

$$y_i = f(x_i, \beta) + \varepsilon_i,$$

where f is a nonlinear function parameterised by β. Iterative algorithms such as the Levenberg–Marquardt method are used to minimise the residual norm:

$$\beta^{(k+1)} = \beta^{(k)} + (J^T J + \lambda I)^{-1} J^T [\mathbf{y} - f(\beta^{(k)})],$$

where J is the Jacobian matrix of partial derivatives of the model with respect to the parameters. The damping term λ transitions between Gauss–Newton and gradient descent steps, enhancing stability.

This approach is particularly useful in fitting thermal profiles across regenerators, where exponential decay or hyperbolic tangent functions better represent the observed physics than polynomials. It is also used in calibrating sensor response curves and interpolating lookup tables for non-ideal gas properties.

8.2.3 Sinusoidal fitting for periodic data

Cryocooler measurements often exhibit sinusoidal oscillations due to periodic compression and expansion of the working gas. These signals are well-described by harmonic models of the form

$$y(t) = A \sin(\omega t + \phi) + y_0,$$

where A is the amplitude, ω the angular frequency, ϕ the phase shift, and y_0 the mean offset. Fitting this model allows extraction of amplitude and phase information, which are directly linked to thermodynamic work and heat transport characteristics.

Phase shift ϕ is especially important in determining the effectiveness of pressure–mass flow coupling, which governs acoustic power delivery and cooling efficiency.

Least-squares estimation can be used with nonlinear solvers or by reparameterising the model into a linear form using sine and cosine basis functions.

8.3 Parametric optimisation

Optimisation techniques enable systematic tuning of cryocooler design parameters to improve performance, subject to practical engineering constraints. Unlike regression, which fits known data, optimisation actively searches for input configurations that minimise or maximise a desired objective function.

8.3.1 Objective functions and constraints

A typical optimisation problem is framed as

$$\text{maximise or minimise} \quad f(x),$$

subject to constraints:

$$g_i(x) \leqslant 0, \quad h_j(x) = 0.$$

For cryocoolers, a common objective is to maximise second-law efficiency or cooling power while minimising input power or vibration. One example cost function is the efficiency:

$$\eta = \frac{\dot{Q}_c}{W_{\text{input}}},$$

where $\dot{Q}_c$ is the cooling load and W_{input} the electrical work. Constraints may include geometric limits (e.g. regenerator length), thermal bounds (e.g. maximum warm end temperature), or dynamic restrictions (e.g. operating frequency).

8.3.2 Gradient-based methods

For smooth objective functions, gradient-based methods such as steepest descent, Newton's method, or Lagrange multipliers are effective. In the presence of equality constraints $g(x) = 0$, the method of Lagrange multipliers solves

$$\nabla f(x^*) + \lambda \nabla g(x^*) = 0,$$

with the constraint enforced directly. These methods are efficient in low-dimensional design spaces and are well-suited for fine-tuning parameters in detailed simulations.

Hessian-based methods may also be used to accelerate convergence when the objective function is well-approximated locally by a quadratic form.

8.3.3 Global search and grid sweeps

In high-dimensional or non-convex problems, global search methods such as brute-force sweeps, Latin hypercube sampling, or Monte Carlo evaluation are employed. These approaches evaluate the objective function over a predefined grid or random sampling of the input space.

For example, a 3D parametric sweep over frequency, charge pressure, and regenerator porosity might reveal global optima in cooling performance or acoustic power. While computationally expensive, such methods provide valuable insight into sensitivity and robustness.

These sweeps are often visualised as contour maps or Pareto surfaces and can serve as a precursor to more refined surrogate-based or Bayesian optimisation techniques.

8.4 Bayesian optimisation

Bayesian optimisation is a global optimisation strategy particularly suited to expensive-to-evaluate or black-box objective functions, where gradients are not available and function evaluations are costly. This makes it well-suited for cryocooler design problems involving detailed simulations or physical testing, where each run may require several hours of computation or experimental set-up.

The method works by constructing a probabilistic surrogate model of the objective function and iteratively selecting new sampling points based on both predicted performance and model uncertainty.

8.4.1 Surrogate models and Gaussian processes

At the core of Bayesian optimisation is the surrogate model: an approximation of the true objective function $f(x)$. A popular choice is the Gaussian process (GP), which defines a distribution over functions:

$$f(x) \sim \mathcal{GP}(\mu(x), k(x, x')),$$

where $\mu(x)$ is the mean function (typically assumed to be zero) and $k(x, x')$ is the covariance kernel, quantifying similarity between input points. Common kernels include the squared exponential (RBF) kernel:

$$k(x, x') = \sigma_f^2 \exp\left(-\frac{(x - x')^2}{2l^2}\right),$$

where σ_f is the signal variance and l is the length scale hyperparameter.

Given a set of observed data points $\mathcal{D} = \{x_i, y_i\}_{i-1}^{n}$, the GP posterior provides both a predicted mean $\mu_n(x)$ and uncertainty $\sigma_n(x)$ at any new point x, allowing the optimisation algorithm to balance exploration and exploitation.

8.4.2 Acquisition functions

The selection of the next sample point is governed by an acquisition function $\alpha(x)$, which quantifies the utility of evaluating the objective at x. A widely used example is the expected improvement (EI), defined as

$$\mathrm{EI}(x) = \mathbb{E}[\max(f(x) - f_{\text{best}}, 0)],$$

where f_{best} is the best observed value so far. Under the GP model, the EI has a closed-form expression:

$$\mathrm{EI}(x) = (m(x) - f_{\text{best}})\Phi(z) + \sigma(x)\phi(z),$$

with

$$z = \frac{m(x) - f_{\text{best}}}{\sigma(x)},$$

where $m(x)$ and $\sigma(x)$ are the predicted mean and standard deviation at x, and Φ, ϕ are the standard normal CDF and PDF respectively.

Other acquisition functions include upper confidence bound (UCB) and probability of improvement (PI), each offering different trade-offs between sampling uncertainty and expected gain.

8.4.3 Cryocooler design example

In cryocooler optimisation, Bayesian methods can be applied to design problems where simulation cost or experimental resource constraints make exhaustive search impractical. For example, to minimise cold-tip temperature T_{cold}, one might define

$$f(x) = T_{\text{cold}}(f_{\text{drive}}, P_{\text{charge}}, \epsilon),$$

where the input vector x consists of compressor frequency f_{drive}, charge pressure P_{charge}, and regenerator porosity ϵ. The surrogate model learns the response surface of T_{cold} and identifies optimal configurations by adaptively sampling high-value regions.

Bayesian optimisation is particularly advantageous when combined with physical constraints or computational models that include stochastic noise or model mismatch. It can also be extended to batch or multi-fidelity settings where coarse simulations are used initially before switching to higher-fidelity evaluations.

8.5 Multi-objective trade-offs

Many cryocooler design problems involve competing objectives that cannot be simultaneously optimised without trade-offs. For instance, increasing cooling power often results in higher vibration, and improving efficiency may require heavier or more complex components.

Multi-objective optimisation formalises these situations by seeking the Pareto front—a set of non-dominated solutions where no objective can be improved without worsening another. The general problem is formulated as

$$\text{minimise} \quad (f_1(x), f_2(x), \ldots, f_m(x)), \qquad x \in \mathcal{X},$$

subject to any constraints on the design space $\mathcal{X}$.

The outcome is a set of Pareto-optimal solutions, often visualised as a curve or surface in objective space. Designers can then select among these based on mission-specific priorities, such as trading off cold-end temperature versus system mass, or efficiency versus lifetime.

Visualisation tools such as parallel coordinate plots or Pareto scatter plots are helpful in understanding the structure of the trade-offs and informing decision-making in multidisciplinary design optimisation (MDO) workflows.

8.6 Summary

This chapter has outlined key techniques for data analysis and optimisation in cryocooler systems. Signal processing methods, including Fourier transforms and digital filtering, enable the extraction of dominant frequencies, phase relationships, and noise characteristics from time-resolved measurements. These tools are essential for diagnosing dynamic performance and validating theoretical models.

Regression and curve fitting approaches provide empirical models for cooling power, temperature profiles, and acoustic behaviour. Linear, nonlinear, and sinusoidal fits are used to interpret trends and infer system parameters from experimental data.

Parametric optimisation frameworks support the systematic tuning of design variables, such as frequency, pressure, and geometry, to enhance cryocooler efficiency, stability, and cooling capacity. Gradient-based methods are suited to local optimisation, while grid sweeps and Bayesian optimisation allow exploration of high-dimensional or expensive design spaces.

Finally, multi-objective optimisation techniques formalise the trade-offs inherent in cryogenic design, offering insight into the balance between performance metrics such as efficiency, vibration, and mass. These tools support rational decision-making in both experimental planning and system-level engineering.

References

Boyd S and Vandenberghe L 2004 *Convex Optimization* (Cambridge: Cambridge University Press)

Bracewell R 2000 *The Fourier Transform and its Applications* (New York: McGraw-Hill)

Brochu E, Cora V M and De Freitas N A 2010 Tutorial on Bayesian optimisation of expensive cost functions arXiv: 1012.2599

Cooley J W and Tukey J W 1965 An algorithm for the machine calculation of complex Fourier series *Math. Comput.* **19** 297–301

Deb K 2001 *Multi-Objective Optimization Using Evolutionary Algorithms* (New York: Wiley)

Jones D R, Schonlau M and Welch W J 1998 Efficient global optimisation of expensive black-box functions *J. Glob. Optim.* **13** 455–92

More J J 1978 The Levenberg–Marquardt algorithm: implementation and theory *Numerical Analysis* Lecture Notes in Mathematics vol 630 *(Berlin: Springer)*

Nocedal J and Wright S J 2006 *Numerical Optimization* (Berlin: Springer)

Oppenheim A V and Schafer R W 2010 *Discrete-Time Signal Processing* (Englewood Cliffs, NJ: Prentice Hall)

Radebaugh R 2000 Development of the pulse tube refrigerator as an efficient and reliable cryocooler *Proc. Inst. Refrig.* **96** 11–29

Rasmussen C E and Williams C K I 2006 *Gaussian Processes for Machine Learning* (Cambridge, MA: MIT Press)

Seber G A F and Lee A J 2012 *Linear Regression Analysis* (New York: Wiley)

Smith S W 1997 *The Scientist and Engineer's Guide to Digital Signal Processing* (San Diego, CA: California Technical Publishing)

Hannah Rana

Chapter 9

Conclusions

This book has explored the mathematical foundations and applied modelling techniques that underpin the design, analysis, and optimisation of cryocoolers. Each chapter has contributed a critical piece to this multidisciplinary puzzle, collectively equipping the reader with a versatile toolkit for addressing challenges in low temperature theory, design, and experiment.

Chapter 1 provided a historical and technical foundation, charting the evolution of cryogenics and the emergence of various cryocooler architectures. It highlighted the critical role cryocoolers play in scientific, aerospace, and quantum technologies, and framed the need for sophisticated mathematical modelling to improve performance and extend system lifetime. Chapter 2 introduced the essential mathematical fundamentals, including complex numbers, phasors, expansion series, and differential equations. These tools form the analytical basis for modelling oscillatory systems, transient behaviour, and coupled physical domains across cryogenic devices. Chapter 3 delved into thermodynamics, with a particular emphasis on the first and second laws, enthalpy flows, and entropy generation. It introduced exergy as a measure of system performance and provided rigorous cycle-level insights into the energy efficiency of cryocoolers. The chapter concluded with detailed treatments of real and ideal thermodynamic cycles, as well as P–V diagram interpretation. Chapter 4 developed harmonic approximations and phasor-based models, demonstrating how sinusoidal behaviour in pressure, flow, and displacement can be analysed using steady-periodic methods. This chapter linked mathematical idealisations to physical intuition, emphasising resonance, impedance, and power transport in Stirling and pulse tube systems.

Chapter 5 addressed numerical modelling, comparing frequency-domain and time-domain approaches. It covered 1D modelling in Sage, finite volume solvers, mesh convergence, and empirical loss models. This chapter emphasised trade-offs between accuracy, computational cost, and modelling fidelity, and introduced techniques for benchmarking simulation outputs. Chapter 6 presented thermoacoustic theory as a

doi:10.1088/978-0-7503-4826-3ch9

unifying framework for describing acoustic power transport and phasing in regenerators and pulse tubes. Rott's functions, boundary layer theory, and worked examples were used to demonstrate how wave-based field models extend beyond lumped parameter methods to resolve spatial distributions and phase-sensitive losses. Chapter 7 focused on hydrodynamic modelling and computational fluid dynamics (CFD). It outlined the Navier–Stokes equations and their adaptations for cryocooler flows, including local thermal non-equilibrium (LTNE) modelling in porous regenerators and turbulence modelling. Realistic case studies illustrated the application of CFD tools in diagnosing flow separation, vortex formation, and distributed losses.

Chapter 8 explored data analysis and optimisation techniques. It introduced signal processing methods for cryocooler diagnostics, regression and curve fitting, and both gradient-based and global optimisation strategies. The chapter concluded with an introduction to Bayesian optimisation and multi-objective tradeoff strategies, demonstrating how machine learning can augment traditional cryogenic design.

Overall, this book integrates thermodynamics, applied mathematics, acoustics, and fluid dynamics into a coherent framework for cryocooler analysis. It balances theoretical depth with practical application, providing both foundational understanding and state-of-the-art modelling strategies. By bridging these domains, it enables engineers, physicists, and researchers to build more efficient, reliable, and innovative cryogenic systems.

Future outlook

As cryocooler technologies evolve to meet the demands of quantum systems, deep-space observatories, and sustainable aerospace, the role of mathematical modelling will only grow in importance. Future developments are expected to integrate real-time optimisation, AI-assisted fault prediction, and advanced multiphysics simulations into the cryocooler design loop. These advances will demand ever more robust and adaptable mathematical methods, building upon the foundations laid out in this book.

www.ingramcontent.com/pod-product-compliance
Ingram Content Group UK Ltd.
Pitfield, Milton Keynes, MK11 3LW, UK
UKHW052037060226
467778UK00001B/6

9 780750 348249